T0280702

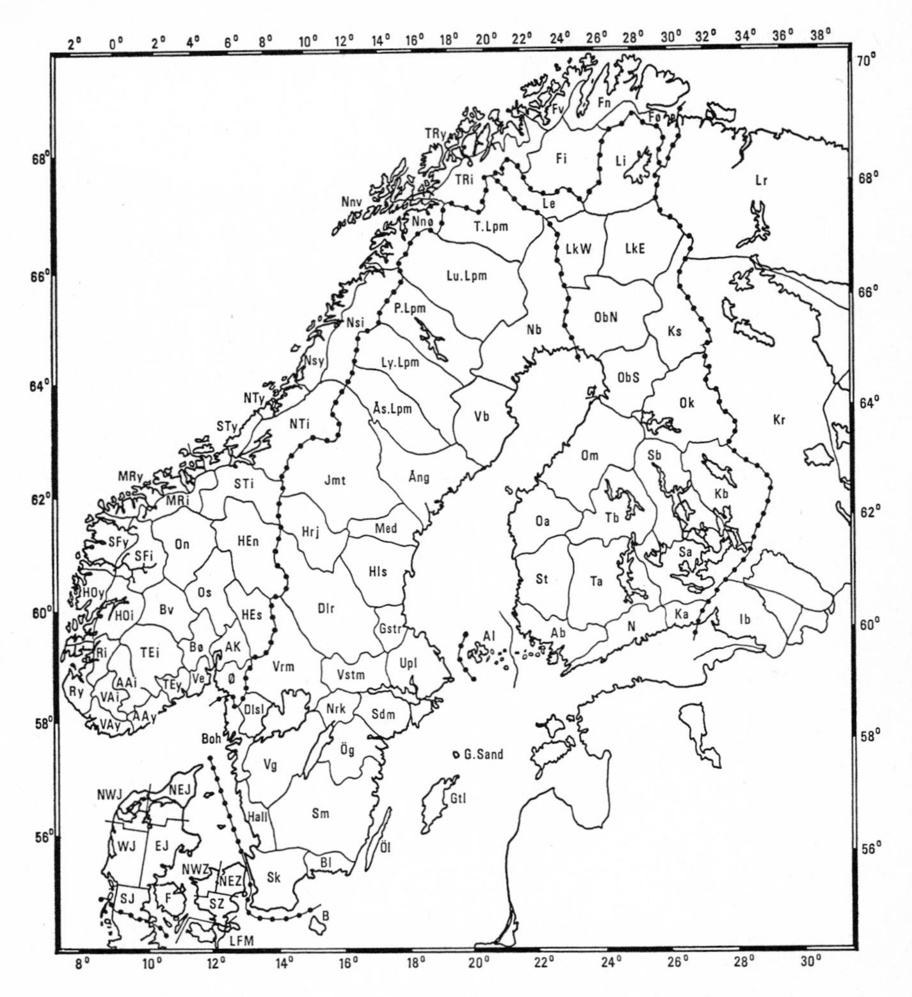

List of abbreviations for the provinces used throughout the text, on the map and in the following tables.

DENMARK

SJ	South Jutland
EJ	East Jutland
WJ	West Jutland
NWJ	North West Jutland
NEJ	North East Jutland
F	Funen
LFM	Lolland, Falster, Møn
SZ	South Zealand
NWZ	North West Zealand
NEZ	North East Zealand
B	Bornholm

SWEDEN

Sk.	Skåne	Vrm.	Värmland
Bl.	Blekinge	Dlr.	Dalarna
Hall.	Halland	Gstr.	Gästrikland
Sm.	Småland	Hls.	Hälsingland
Öl.	Öland	Med.	Medelpad
Gtl.	Gotland	Hrj.	Härjedalen
G. Sand.	Gotska Sandön	Jmt.	Jämtland
Ög.	Östergötland	Ång.	Ångermanland
Vg.	Västergötland	Vb.	Västerbotten
Boh.	Bohuslän	Nb.	Norrbotten
Dlsl.	Dalsland	Ås. Lpm.	Åsele Lappmark
Nrk.	Närke	Ly. Lpm.	Lycksele Lappmark
Sdm.	Södermanland	P. Lpm.	Pite Lappmark
Upl.	Uppland	Lu. Lpm.	Lule Lappmark
Vstm.	Västmanland	T. Lpm.	Torne Lappmark

NORWAY

Ø	Østfold	HO	Hordaland
AK	Akershus	SF	Sogn og Fjordane
HE	Hedmark	MR	Møre og Romsdal
O	Opland	ST	Sør-Trøndelag
B	Buskerud	NT	Nord-Trøndelag
VE	Vestfold	Ns	southern Nordland
TE	Telemark	Nn	northern Nordland
AA	Aust-Agder	TR	Troms
VA	Vest-Agder	F	Finnmark
R	Rogaland		

n northern s southern ø eastern v western y outer i inner

FINLAND

Al	Alandia	Kb	Karelia borealis
Ab	Regio aboensis	Om	Ostrobottnia media
N	Nylandia	Ok	Ostrobottnia kajanensis
Ka	Karelia australis	ObS	Ostrobottnia borealis, S part
St	Satakunta	ObN	Ostrobottnia borealis, N part
Ta	Tavastia australis	Ks	Kuusamo
Sa	Savonia australis	LkW	Lapponia kemensis, W part
Oa	Ostrobottnia australis	LkE	Lapponia kemensis, E part
Tb	Tavastia borealis	Li	Lapponia inarensis
Sb	Savonia borealis	Le	Lapponia enontekiensis

USSR

Vib Regio Viburgensis Kr Karelia rossica Lr Lapponia rossica

FAUNA ENTOMOLOGICA SCANDINAVICA
Volume 11 1982

The Aphidoidea (Hemiptera) of Fennoscandia and Denmark. II

The family Drepanosiphidae

by

Ole E. Heie

SCANDINAVIAN SCIENCE PRESS LTD.
Klampenborg . Denmark

Fauna entomologica scandinavica
is edited by »Societas entomologica scandinavica«

World list abbreviation
Fauna ent. scand.

Printed by
Vinderup Bogtrykkeri A/S
7830 Vinderup, Denmark

ISBN 87-87491-44-3
ISSN 0106-8377

Contents

Plates 1 and 2 are arranged between pp. 16 and 17.
Plates 3 and 4 are arranged between pp. 32 and 33.

Preface

Volume 11 of "Fauna entomologica scandinavica" is the second of four planned volumes dealing with the superfamily Aphidoidea of the infraorder Aphidodea Mordvilko (= Aphidoidea of Börner) within the order Homoptera, suborder Sternorrhyncha.

The first volume appeared as volume 9 of "Fauna entomologica scandinavica" (1980). It comprised introductury chapters to the superfamily, keys to families, subfamilies, and tribes, and descriptions and keys to Scandinavian genera and species of Mindaridae, Hormaphididae, Thelaxidae, Anoeciidae, and Pemphigidae.

The present second volume deals with the Drepanosiphidae, including the subfamilies Drepanosiphinae, Phyllaphidinae, and Chaitophorinae.

The third volume will deal with part of the Aphididae, viz. Pterocommatinae and Aphidinae: Aphidini. The fourth volume will deal with Aphidinae: Macrosiphini and Lachnidae.

A zoogeographical catalogue is given in each of the four volumes, references are also given in each volume, but those listed in the first volume do not reappear in the following volumes. Each volume has its own index of entomological names. The last volume is concluded with an index listing all entomological names in the four volumes and a host plant index.

The Scandinavian species are provided with numbers continuing throughout the four volumes, but each volume is separately paged.

The geographical terminology is the same as in the first volume. The following abbreviations are used in the descriptions and/or figures:

ant. segm.: antennal segment; abd. segm.: abdominal segment; apt. viv.: apterous viviparous (and parthenogenetic) female; al. viv: alate viviparous (and parthenogenetic) female; ovip.: oviparous (not parthenogenetic) female; Survey: Eastop, V. F. & Hille Ris Lambers, D., 1976: Survey of the World's aphids. – Junk, the Hague, 573 pp. (reference for synonyms); IIIbd.: basal diameter of ant. segm. III (Fig. 57 in volume 9); VIa: basal part of ant. segm. VI (from base to distal margin of primary rhinarium) (Fig. 56 in vol. 9); VIb: processus terminalis, the ultimate thinner part of ant. segm. VI; 2sht.: second segment of hind tarsus (without claws).

Further information is given in the preface and introductury chapters of the first volume (vol. 9).

Acknowledgements

I wish to thank the colleagues, institutions, and foundations mentioned in the first volume (p. 12), and also the Natural History Museum, Århus, for laboratory facilities and assistance, the Royal Veterinary and Agricultural University, Zoological Institute, Copenhagen, and Mrs. J. Cole, Dr. L. R. Taylor and other members of the staff at the Rothamsted Experimental Station, Harpenden, England, for cooperation in connection with records from the Rothamsted Insect Survey suction trap at Tåstrup, Zealand, Mr. M. Gissel Nielsen, Århus, for loan and gift of material, and Mr. J. Balle Hansen, Skive, for linguistic criticism.

The Danish Science Research Council, the Carlsberg Foundation, and the Zoological Museum of Copenhagen are thanked for financial support or other facilities.

Family Drepanosiphidae

All instars of all morphs of Scandinavian species have large compound eyes, with or without ocular tubercles (triommatidia). Processus terminalis usually longer than 0.5 × basal part of ultimate antennal segment, sometimes very long. Secondary rhinaria transverse oval or almost circular, usually present only on ant. segm. III in females, often absent from apterous females. Empodial hairs usually more or less flattened. Media of fore wing with one or two forks. Hind wing usually with two oblique veins. Wings roof-like in repose in Scandinavian species. Siphunculi present from birth (except in some non-Scandinavian genera of Drepanosiphinae), pore-shaped to very long, up to about 0.3 × body length, usually short and stump-shaped, sometimes reticulate (Fig. 1–4). Cauda broadly rounded (semicircular) or knobbed (Figs. 36, 37, 311, 312). Anal plate usually emarginate (Fig. 16) or bilobed (Fig. 8).

All species are monoecious. The Scandinavian species are monophagous or oligophagous on deciduous trees, monocotyledones or (*Therioaphis*) on herbaceous Leguminosae. The family apparently predominated in the beginning of the Tertiary period. Since then it has been standing down in favour of the family Aphididae, but it is still rather rich in species.

The family Drepanosiphidae as here understood corresponds to the families Callaphididae + Chaitophoridae of Börner (1952) and the *Phyllaphis*- or *Callaphis*-group + the *Chaitophorus*-group (subfamilies, tribes) of several other authors. This family concept is the same as that of Bodenheimer & Swirski (1957), who called the family Callipteridae. For argumentation, comments on nomenclature, and subdivision, see Heie (1980: 17–19).

Key to subfamilies of Drepanosiphidae

Apterous and alate viviparous females

1 Siphunculi (of Scandinavian species) about 0.2 × body or longer, cylindrical or slightly swollen (Fig. 1). With three rudimentary gonapophyses (Fig. 7). .... **Drepanosiphinae** (p. 12)

– Siphunculi shorter than 0.2 × body. With 2 or 4 rudimentary gonapophyses. 2

2 (1) With 2 rudimentary gonapophyses (Fig. 8). Basal part of segment II of rostrum usually with sclerotized wishbone-shaped arch (Figs. 5, 6). Siphunculi never reticulate (Figs. 2, 3) ...... **Phyllaphidinae** (p. 18)

– With 4 rudimentary gonapophyses (Fig. 9). Basal part of segment II of rostrum without sclerotized arch. Siphunculi sometimes reticulate (Fig. 4). .... **Chaitophorinae** (p. 106)

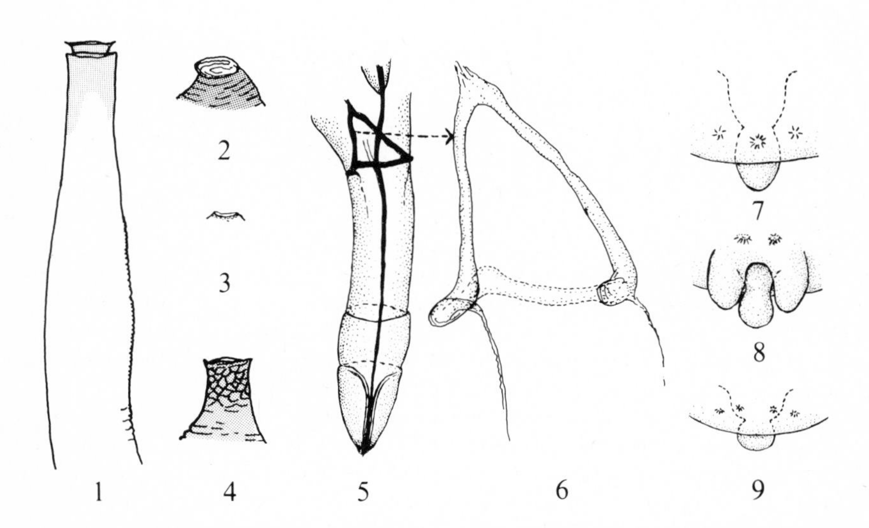

Figs. 1–4. Siphunculi of genera belonging to different subgroups of Drepanosiphidae. – 1: *Drepanosiphum* (Drepanosiphinae); 2: *Callaphis* (Phyllaphidinae); 3: *Phyllaphis* (Phyllaphidinae); 4: *Periphyllus* (Chaitophorinae).

Figs. 5, 6. Rostrum of *Callaphis juglandis* (Goeze) (Phyllaphidinae), showing the basal wishbone-shaped sclerotization.

Figs. 7–9. Posterior part of abdomen, ventral view. – 7: *Drepanosiphum* (Drepanosiphinae) with three rudimentary gonapophyses; 8: *Thripsaphis* (Phyllaphidinae) with two gonapophyses; 9: *Chaitophorus* (Chaitophorinae) with four gonapophyses.

SUBFAMILY DREPANOSIPHINAE

All viviparous females, including fundatrices, alate (except in some non-Scandinavian species). Antennae 6-segmented. Not all accessory rhinaria on ant. segm. VI are placed close to the primary rhinarium (except in *Yamatocallis* Matsumura from E Asia). but form an irregular row (Fig. 11). Segment II of rostrum without sclerotized wishbone-shaped arch at base (Fig. 10). Fore leg in most genera, including *Drepanosiphum,* with thickened and prolonged femur (Fig. 12). Empodial hairs spatulate (Fig. 13). Siphunculi of various lengths and shapes. Three rudimentary gonapophyses present (Figs. 7, 15), the middle one may be subdivided in some non-Scandinavian species. Cauda with a basal constriction, either knobbed or elongate (Figs. 14, 16). Anal plate of viviparous females slightly emarginate, in *Drepanosiphum* almost semicircular in dorsal view, in some other genera bilobed; anal plate of oviparous females rounded.

The hosts are woody plants and herbs belonging to several families, e. g. Aceraceae, Fagaceae, and Gramineae. All Scandinavian species live on *Acer.* The aphids are not visited by ants. In Scandinavia only one genus belonging to the tribe Drepanosiphini (Heie, 1980: 18).

Genus *Drepanosiphum* Koch, 1855

Drepanosiphum Koch, 1855: 201.

Type-species: *Aphis platanoidis* Schrank, 1801.

Survey: 179.

Antennae longer than body; processus terminalis several times as long as VIa; viviparous female with transverse oval secondary rhinaria on basal part of segm. III, male with subcircular secondary rhinaria on segm. III–V. Eyes prominent. Frons with well developed, diverging lateral tubercles. Siphunculi long, 0.20–0.33 × body, slightly swollen or almost cylindrical, smooth, with a ringlike contriction below the flange (Fig. 1). Cauda knobbed, with a more or less distinct constriction.

The oviparous female is the only apterous morph. The posterior end of the abdomen is prolonged like a cone and functions as an ovipositor.

All species live on the undersides of leaves of *Acer* spp., often in large numbers, but never in dense colonies. The reproduction is low in mid summer, when numerous alatae – but, at most, a few nymphs – occur evenly spaced all over the lower surface of the leaves (Fig. 17A) (Dixon 1976). Leaves in shade are preferred. The individuals sit just as close to each other so to reach the neighbours with the antennae and become alert if an individual is being disturbed by an enemy. The whole leaf can be left rapidly by all the aphids. They stretch out the fore legs almost simultaneously so that the stylets are redrawn from the plant tissue and the body thrown away from the leaf. The aphids prefer to settle on the minor veins of large leaves (Fig. 17A), but on the larger veins of small leaves (Fig. 17B). Individuals feeding on large leaves are unable to reach the sieve tubes in the thicker part of a main vein as their stylets are too short (Dixon & Logan 1973). In strong wind they aggregate close to the protruding basal parts of the large veins to get shelter from the effect of the leaves brushing against one another (Fig. 17C). As soon as the wind has abated they spread again (Fig. 17D) so that competition is reduced (Dixon & McKay 1970).

The sexuales are produced in autumn and move to the branches and the trunk, where the eggs are laid in bark crevices and behind buds.

There are 8 species in the world, 3 in Scandinavia. Two more species not yet found in Scandinavia are included in the key and the descriptions below. The key characters are taken from Hille Ris Lambers (1971).

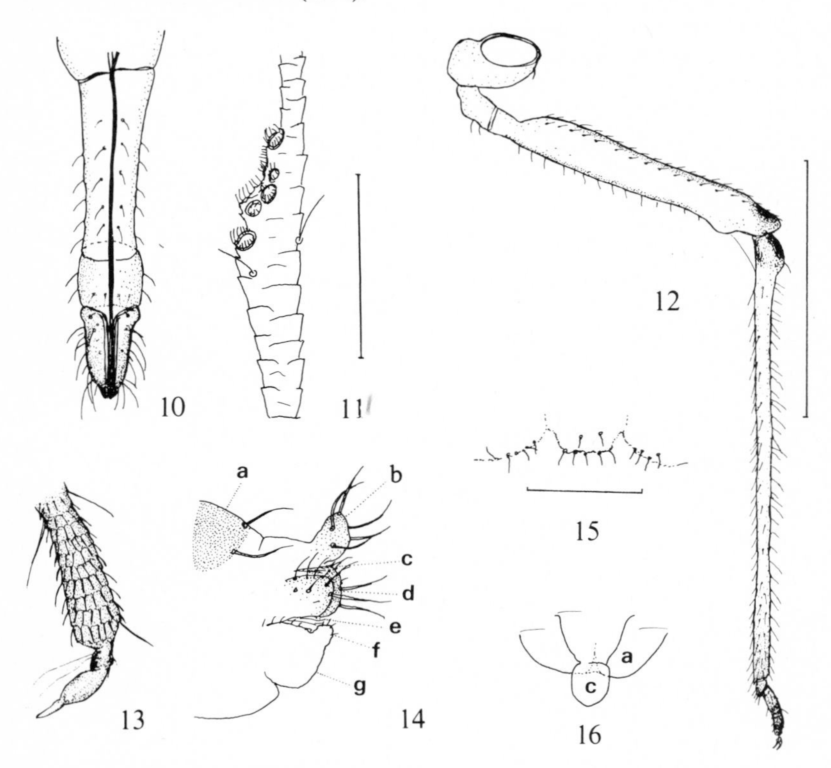

Figs. 10–16. *Drepanosiphum platanoidis* (Schrk.), al. viv. – 10: rostrum; 11: basal part of ant. segm. VI with accessory rhinaria arranged in irregular row near the larger primary rhinarium; 12: fore leg; 13: apex of hind tarsus with spatulate empodial hairs under the claws; 14: posterior part of abdomen from left side; a = abd. tergite VIII, b = cauda, c = anus, d = anal plate, e = genital aperture, f = rudimentary gonaphyses, g = genital plate; 15: rudimentary gonaphyses (ventral view); 16: cauda (c) and anal plate (a) in outline, dorsal view. (Scales 1 mm for 12, 0.1 mm for 11 and 15).

Key to species of *Drepanosiphum*

Alate viviparous females
(Apterous viviparous females do not occur)

1 Fore wing with a dark apical spot, and a dark spot at the distal end of the pterostigma (Fig. 19). .... 56. *aceris* Koch
– Fore wing without dark apical spot, but sometimes with a dark spot at the distal end of the pterostigma. 2
2 (1) Pterostigma with a dark spot distally (Fig. 20). Fore femur without dark ventral stripe, Abdomen with pleurospinal bars on tegites IV and V. Wings often very short. .... *dixoni* Hille Ris Lambers
– Pterostigma without dark spot distally, only more or less distinct, brownish band posteriorly (Fig. 18). Fore femur often with a dark ventral stripe. Abdomen without bars, or with two or more bars. Wings never very short. 3
3 (2) Ant. segm. III paler at base than at apex. Apical segment of rostrum 0.14 mm or longer. .... 57. *platanoidis* (Schrank)
– Ant. segm. III with an area near base darker than the rest, or entirely pale. Apical segment of rostrum rarely longer than 0.13 mm. 4
4 (3) Base of ant. segm. III with darker area. Fore femur much thicker than hind femur, with distinct blackish ventral stripe (Fig. 25). Only very dark specimens have blackish spots near bases of siphunculi. .... *oregonensis* Granovsky
– Base of ant. segm. III not darker than the rest of the segment. Fore femur often hardly thicker than hind femur, rarely with

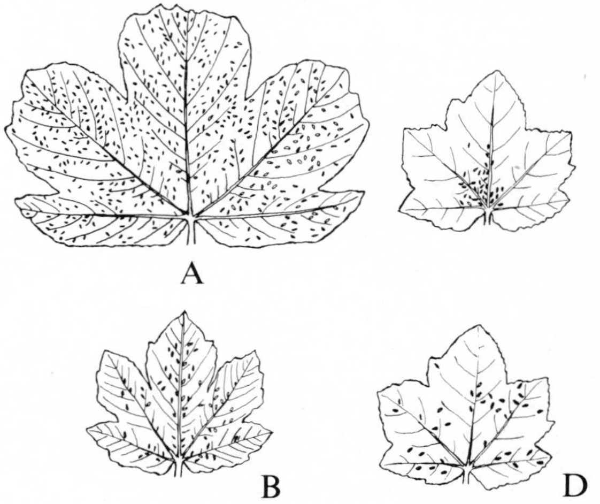

A

B

D

Fig. 17. Distribution of adult *Drepanosiphum platanoidis* (Schrk.) on the undersides of sycamore leaves *(Acer pseudoplatanus)* relative to the large veins on a large leaf (A), and on a small leaf (B), during a strong wind (C), and after the wind has abated (D). (After Dixon, redrawn).

indication of blackish ventral stripe. Black spots usually present near bases of siphunculi, also in specimens with rather pale thorax (Fig. 28). .... 55. *acerinum* (Walker)

55. ***Drepanosiphum acerinum*** (Walker, 1848)
Figs. 21, 23, 28.

Aphis acerina Walker, 1848 b: 254. – Survey: 179.

Alate viviparous female. Yellow or whitish yellow, with darker thorax. Abdomen with 1–2 dark cross bars in front of the siphunculi and usually with small dark spots near bases of siphunculi (Fig. 28). Dorsal body hairs fine, pointed, only 0.07–0.10 mm. Antenna about twice as long as body; apices of segments brownish; base of segm. III without dark area, but not paler than apical part; segm. III with 10–13 oval rhinaria on basal 40% (Fig. 23). Apical segm. of rostrum shorter than 0.13 mm, with about 5–6 accessory hairs. Fore femur less thickened than in the other species, hardly thicker than hind femur, and not or hardly thicker than siphunculus, rarely with pigmentation ventrally. Siphunculus with dark apex or entirely dark, sometimes blackish, about 0.3×body, only little swollen (Fig. 21). Otherwise rather similar to *platanoidis*. 2.1–3.3 mm.

Oviparous female. According to the original description similar to *platanoidis*.

Alate male. Similar to the alate female, but darker.

Distribution. In Denmark rare, found in NEJ (Ålborg, 1 nymph), NWJ (Klitmøller, 1 al.viv), and EJ (Randbøldal, 1 al. viv.); not in Sweden, Norway, and Finland. – British Isles (local), the Netherlands, Germany (rare, not in N Germany), Poland (not common, not in the Baltic region), Czechoslovakia, and Portugal.

Biology. The species lives on *Acer pseudoplatanus*, usually in shade.

56. ***Drepanosiphum aceris*** Koch, 1855
Plate 2:1. Figs. 19, 24, 29, 30.

Drepanosiphum aceris Koch, 1855: 202. – Survey: 179.

Alate viviparous female. Abdomen yellowish with two or more dark cross bars and dark spots in front of the siphuncular bases (Figs. 29, 30). Head and thorax a little darker than abdomen. Dorsal body hairs up to 0.13 mm, more than 2.5×IIIbd., pointed. Antenna nearly twice as long as body; segm. III, IV, and V with dark apices; segm. III with 11–16 oval rhinaria on basal 40–45% (Fig. 24). Apical segm. of rostrum 0.15–0.17 mm, as long as or longer than 2sht., with about 14–19 accessory hairs. Fore wing with a dark apical spot and another dark spot at the distal end of pterostigma (Fig. 19). Siphunculus with dark apex or entirely dark, about 0.30×body. 3.2–3.9 mm.

Oviparous female. Abdomen with dark cross bars and spots. Antennae without secondary rhinaria. Hind tibiae hardly swollen, with scent plaques on basal $^{2}/_{3}$.

Alate male. Abdomen with several dark cross bars. Secondary rhinaria: III: about 100–115, IV: 50 or more, V: 16–21.

Distribution. In Denmark common and widespread; in Sweden found in Ög. and Upl.; not in Norway and Finland. – Europe, e.g. in Great Britain, Germany, Poland, NW Russia, Switzerland, Czechoslovakia, and Hungary; Asia: the Caucasus region.

Biology. The host is *Acer campestre.*

Drepanosiphum dixoni Hille Ris Lambers, 1971
Fig. 20.

Drepanosiphum dixoni Hille Ris Lambers, 1971: 72. – Survey: 179.

Alate viviparous female. Abdomen greenish white with two broad, almost black cross bars in front of siphunculi and dark antesiphuncular sclerites. Antennae, except segm. I, II, and sometimes base of III, and legs, except knees, pale. Dorsal body hairs 0.06–0.07 mm. Antenna twice as long as body or longer; segm. III with 9–20 oval secondary rhinaria on basal 40–45%; hairs on segm. III 0.25–0.33 × IIIbd. Apical segm. of rostrum 0.14–0.17 mm, a little longer than 2sht., with 8–12 accessory hairs. Fore wing with a dark spot at the distal end of pterostigma, but without dark apical spot (Fig. 20). Wings often very short. Fore femur without dark ventral stripe. Siphunculus usually pale with brown apex and faintly pigmented base, rarely entirely black, about 0.30 × body, only slightly swollen. Cauda more or less constricted. 2.5–3.3 mm.

Oviparous female. Similar to *platanoidis,* but with small ocelli and 2–9 small, circular secondary rhinaria on ant. segm. III.

Alate male. Similar to the alate female. Slender. Secondary rhinaria: III: 70–90, IV: 30–40, V: 10–12. About 2.4 mm.

Distribution. Not in Scandinavia. – England (rare; Hertford, Cambridge), the Netherlands, Yugoslavia; probably widespread in Europe, but overlooked.

Biology. The species lives on *Acer campestre.* Brachypterous alatae are common in summer.

Drepanosiphum oregonensis Granovsky, 1939
Fig. 25.

Drepanosiphum oregonensis Granovsky, 1939: 143.
Drepanosiphum zimmermanni Börner, 1940: 2.
Survey: 180.

Alate viviparous female. Abdomen pale green; thorax orange-brown; sometimes with blackish mesosternum and black spots on abdomen near bases of siphunculi. Base of ant. segm. III with an area darker than the rest of the segment. Apical segment of rostrum rarely longer than 0.13 mm. Fore femur with blackish ventral stripe (Fig. 25). Otherwise rather similar to *platanoidis.*

Oviparous female. Olive, with dark cross bars.

Alate male. Head and thorax black; abdomen with black cross bars.

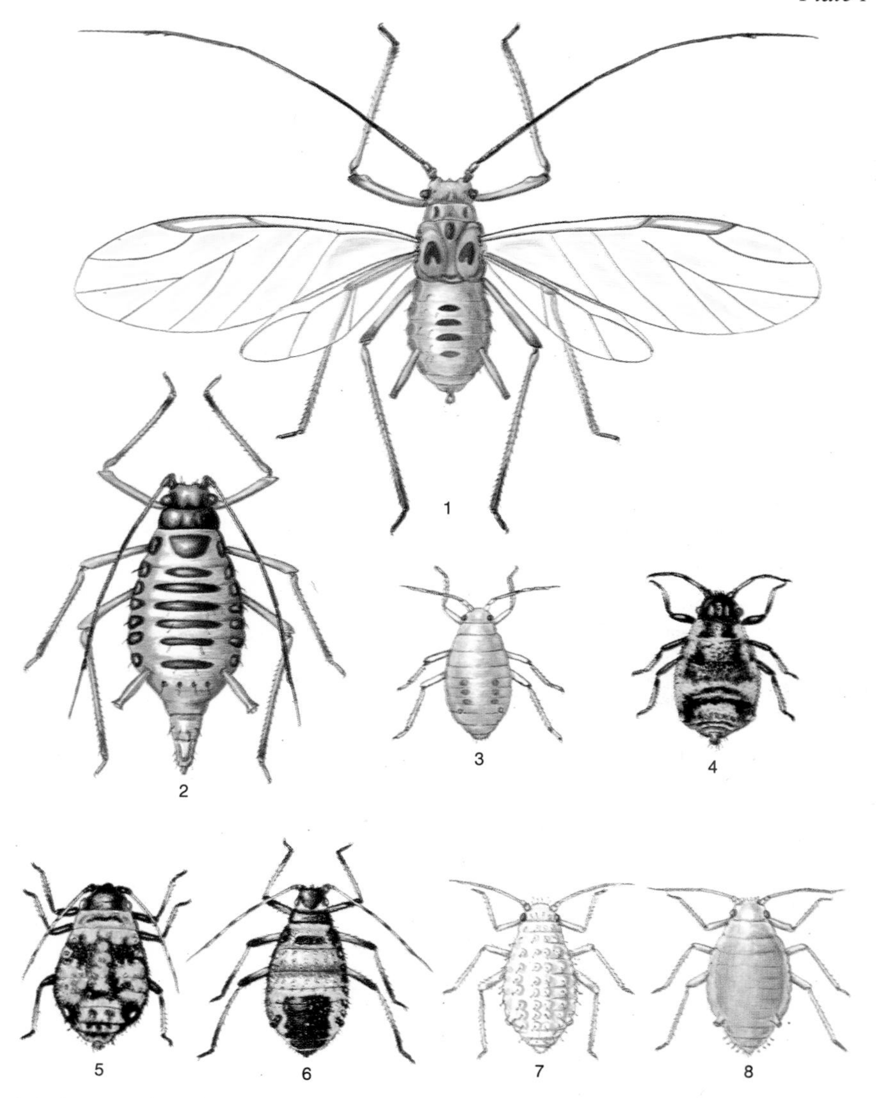

1–2. *Drepanosiphum platanoidis* (Schr.), al. viv. (1) and ovip. female (2). – 3. *Phyllaphis fagi* (L.), apt. viv. (wax wool removed). – 4. *Iziphya leegei* Börn., apt. viv. – 5. *I. bufo* (Wlk.), apt. viv. – 6. *Callipterinella tuberculata* (v. Heyd.), apt. viv. – 7. *Betulaphis quadrituberculata* (Kalt.), apt. viv. – 8. *B. brevipilosa* Börn., apt. viv (10 ×).

Plate 2

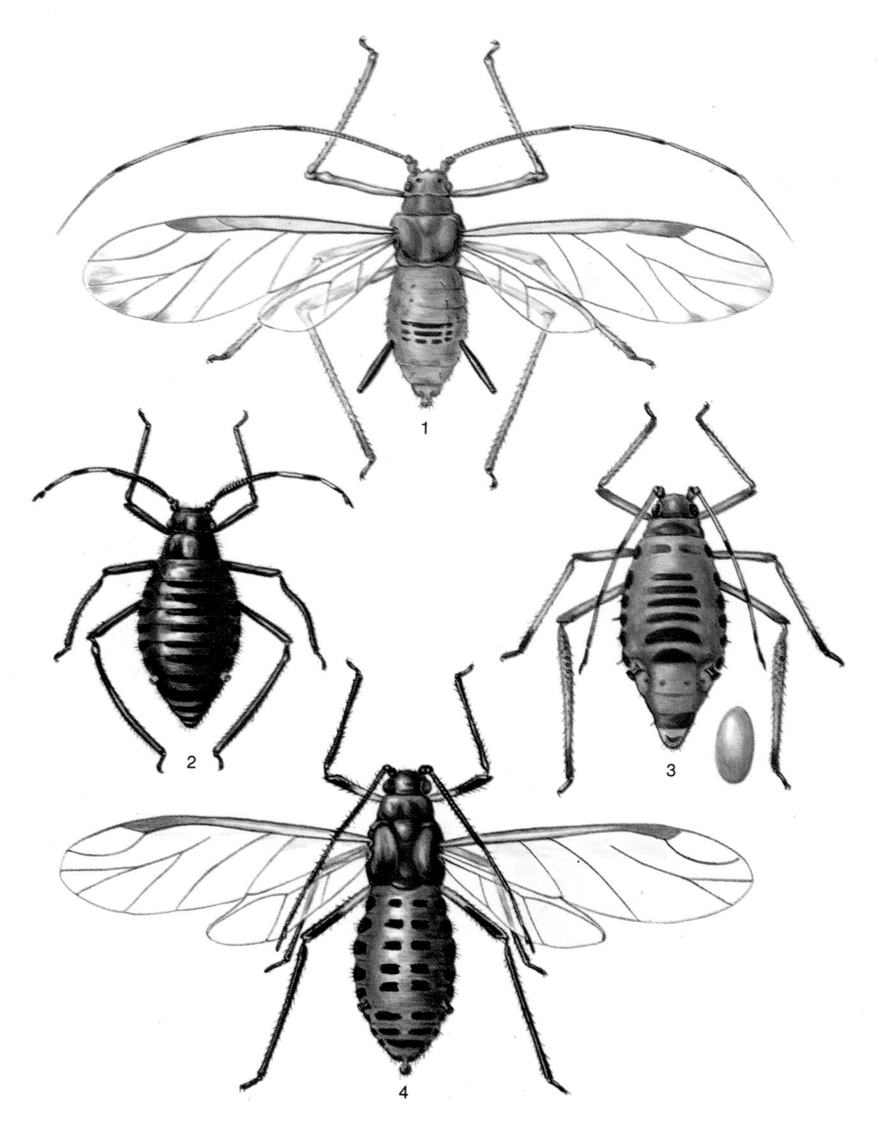

1. *Drepanosiphum aceris* Koch, al. viv. – 2. *Symydobius oblongus* (v. Heyd.), apt. viv. – 3. *Euceraphis punctipennis* (Zett.), ovip. female and egg. – 4. *Clethrobius comes* (Wlk.), al. viv. (8 ×).

Distribution. Not in Scandinavia. – Europe, e.g. Germany, Portugal, and Crimea, but not known from the British Isles and Poland; in N Germany only found in Rostock Botanical Garden (F. P. Müller); N America.

Biology. The species lives on *Acer monspessulanum, A. opalus,* and *A. macrophyllum.*

57. ***Drepanosiphum platanoidis*** (Schrank, 1801)
Plate 1:1,2. Figs. 1, 7, 10–18, 22, 26, 27.

Aphis platanoidis Schrank, 1801: 112. – Survey: 180.

Alate viviparous female. Light green. Head and thorax yellowish or brownish, sometimes dark. Abdomen with 5–6 dark dorsal cross bars and marginal sclerites, including one in front of each of the siphunculi (Fig. 27), or – in early generations –

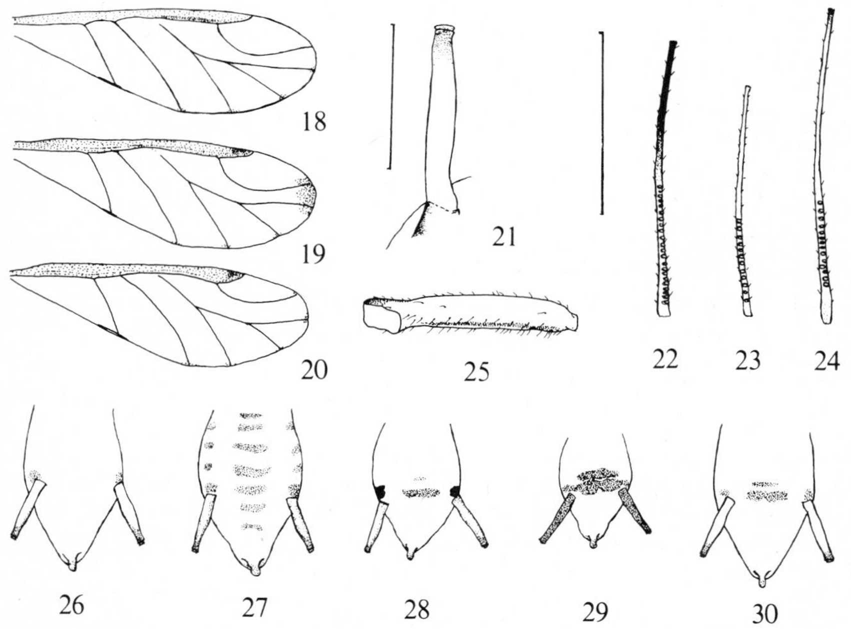

Figs. 18–20. Fore wings. – 18: *Drepanosiphum platanoidis* (Schrk.); 19: *D. aceris* Koch; 20: *D. dixoni* H. R. L. (19 and 20 after Hille Ris Lambers, redrawn).

Fig. 21. Siphunculus of lectotype of *D. acerinum* (Wlk.). (After Doncaster, redrawn).

Figs. 22–24. Ant. segm. III of al. viv. – 22: *D. platanoidis;* 23: *D. acerinum;* 24: *D. aceris.* (Scale 1 mm).

Fig. 25. Fore femur of *D. oregonensis* Granovsky. (After Hille Ris Lambers, redrawn).

Figs. 26–30. Abdomen of al. viv. *Drepanosiphum* spp., dorsal view, hairs omitted. – 26, 27: *platanoidis;* 28: *acerinum;* 29, 30: *aceris.* (26–29 partly after Szelegiewicz).

without any cross bars and sclerites at all (Fig. 26). Dorsal body hairs 0.17–0.18 mm, about 3 × IIIbd., pointed. Antennae 1.2–1.3 × body, pale brownish or dark with paler basal half of segm. III and paler processus terminalis; segm. III with 17–26 secondary rhinaria on line on basal 40–55% (Fig. 22); antennal hairs short, about 0.5 × IIIbd. Apical segm. of rostrum 0.14–0.17 mm, shorter than 2sht., with about 12–15 accessory hairs (Fig. 10). Legs with knees, apices of tibiae, and tarsi (Fig. 12), dark. Fore wing without dark spots at apex or in pterostigma, but pterostigma with two brown longitudinal stripes, one along fore edge of wing and one along radius, both continuing almost to wing base. Siphunculus about 0.20 × body, pale with brown apex. Cauda pale or brownish, constricted, with 5 hairs. 4.0–4.3 mm.

Oviparous female. Brown. Abdominal segments in front of the siphunculi with cross bars and marginal sclerites. Antennae without secondary rhinaria. Hind tibiae only little swollen in basal half, with several scent plaques. Posterior part of abdomen prolonged. Cauda rounded, not constricted.

Alate male. Slender. Abdomen with dark cross bars in front of and behind the siphunculi. Antennae entirely dark; secondary rhinaria almost circular, on III: about 125, IV: 40–50, V: 5–26.

Distribution. In Denmark very common and widespread; common in southern Sweden, from Sk. north to Upl. and Vstm.; common in southern Norway, north to HOy, HOi, and Os; in Finland only found in Helsinki Botanical Garden. – All over Europe, south to Portugal and Spain, including the British Isles, N Germany, Poland, NW & W Russia, and the Faroes; C Asia and Turkey; N Africa (Atlas Mountains); Tasmania, New Zealand; widespread in the USA and in British Columbia, Canada.

Biology. The species lives on *Acer pseudoplatanus* and several other *Acer* spp., though seldom on *A. campestre*. In Denmark viviparous females occur from May till October.

Note. *Acer platanoides* is according to Börner (1952) one of the hosts of *D. platanoidis*. I have not seen the aphid on *A. platanoides* myself, only on *A. pseudoplatanus*, which is also much more common than *A. platanoides* in Denmark. In Sweden *A. platanoides* is more common, going farther north than *A. pseudoplatanus*, according to Lid (1963) north to Dalarna and Ångermanland, whereas *A. pseudoplatanus* is grown or occurs ferally only north to Uppland. No records from *A. platanoides* are given by Ossiannilsson (1959). Apart from a few records from *A. campestre*, *A. monspessulanum*, and *A. pennsylvanicus* in Botanical Gardens in Lund and Uppsala, Ossiannilsson found the aphids on *A. pseudoplatanus*, and the northern limit of *platanoidis* coincides with the northern limit of this tree.

SUBFAMILY PHYLLAPHIDINAE

All viviparous females, including fundatrices, are alate in some species; in some other species, with both apterous and alate viviparous females, fundatrices may be apterous or alate (alate fundatrices do not occur in most other aphids). Altogether the differences between the various morphs and generations are small. The oviparous female is apterous and similar to the apterous viviparous female in species where this morph occurs, except for the thickened hind tibiae with scent plaques and often also for the shape of the anal plate (Fig. 92) and the presence of two ventral facetted wax gland plates (below the siphunculi) (Figs. 52, 195).

Frons often with 3–5 tubercles. Antennae 6-segmented in most species. Processus terminalis from very short to extremely long. Primary rhinaria in Scandinavian genera (except *Monaphis*) surrounded by short hairs. Basal part of segment II of rostrum supported by a sclerotized wishbone-shaped arch (Fig. 5) (except in *Monaphis*). Legs generally rather long and slender, often developed for jumping, with fore coxae (Figs. 31, 32) or fore and middle femora enlarged (Figs. 185, 186, 210, 213). Tarsi and distal parts of tibiae sometimes with transverse rows of spinules. First tarsal segments with 5–9 hairs. Empodial hairs usually spatulate (Fig. 146), seldom simple. Media with two forks; radial sector sometimes rudimentary or absent. Wings often with blackish or brownish pigmentation. Abdomen often with dorsal and marginal tubercles or processes, which in some species are large and finger-shaped (Figs. 101, 112, 119–121), but low and wart-like in most species. Siphunculi usually short, stump- og pore-shaped, never reticulate. Cauda semicircular or knobbed, seldom (*Betulaphis*) broadly triangular (or conical). Anal plate semicircular, emarginate, or bilobed. With 2 rudimentary gonapophyses (Figs. 8, 45).

All species are monoecious, most of them holocyclic and living from spring till autumn on deciduous trees or bushes. Some species feed on herbaceous plants: Saltusaphidini on monocotyledones, *Therioaphis* on herbaceous Leguminosae.

The subfamily Phyllaphidinae corresponds to the family Callaphididae of Börner (1952) (Drepanosiphidae, Phyllaphididae, Callipteridae of other authors, see Heie, 1980: 19), minus Börner's tribe Drepanosiphonini. Börner (1952) and Börner & Heinze (1957) subdivided Callaphididae into four subfamilies, whereas I subdivide Phyllaphidinae into two tribes, one of them (Phyllaphidini) subdivided into three subtribes corresponding to three of Börner's subfamilies.

Key to tribes of Phyllaphidinae

1 Ocular tubercles present (Fig. 122). Fore femur not thicker than hind femur. First tarsal segments sometimes with more than 5 hairs. .... **Phyllaphidini** (p. 20)

– Ocular tubercles absent (Fig. 156). Fore femur sometimes much thicker than hind femur. First tarsal segments with 5 hairs. ...... **Saltusaphidini** (p. 74)

TRIBE PHYLLAPHIDINI

Viviparous females alate or apterous, sometimes all alate. Compound eyes with ocular tubercles (triommatidia). Frons often with lateral tubercles. Accessory rhinaria on ant. segm. VI placed close to the primary rhinarium (Fig. 79). Legs normal (Fig. 33), or fore legs with enlarged coxae (Figs. 31, 32). First tarsal segments sometimes with more than 5 hairs. Empodial hairs usually spatulate. Anal plate of oviparae rounded, also in species with bilobed anal plate in viviparous females (Figs. 91, 92). Pronotum of all instars not fused with the head, or only partly so in the first instar nymphs of some genera (subtribe Callaphidina).

The hosts are deciduous trees (or bushes) or herbaceous Leguminosae. Some species are visited by ants.

The tribe is subdivided into three subtribes corresponding to the subfamilies of Börner's classification (1952). Two keys are given below, one to the subtribes and one to the genera. The former is partly based on characters of first instar nymphs in accordance with Quednau (1954). The latter key is recommended for determination of adults.

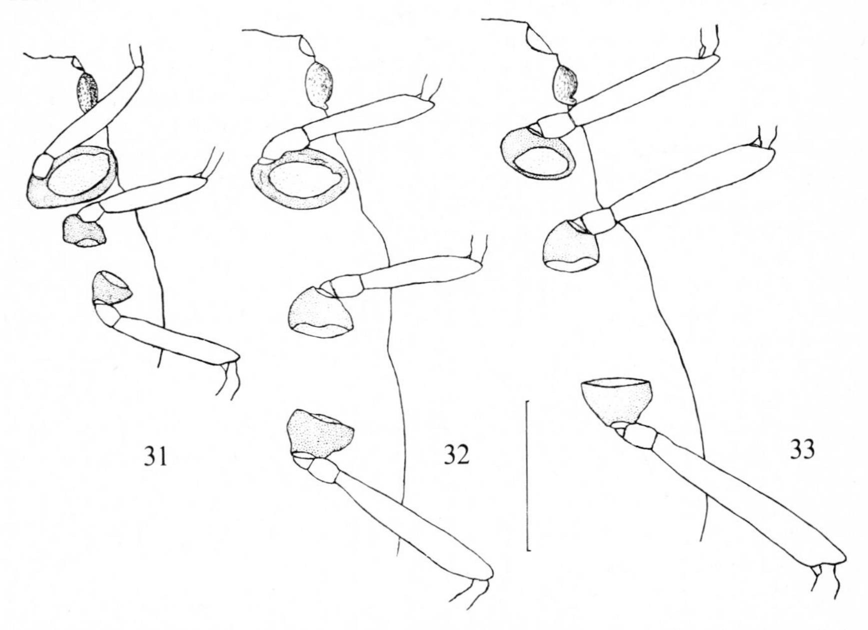

Figs. 31–33. Anterior part of various Phyllaphidini in ventral view; coxae dotted; rostrum, hairs, pigmentation, and segmental borders, omitted. – 31: *Therioaphis;* 32: *Eucallipterus;* 33: *Symydobius.* (Scale 0.5 mm).

Key to subtribes of Phyllaphidini

1 Fore coxae of adults 1.75–3 times as broad as middle coxae (Fig. 31). .... **Therioaphidina** (p. 67)
– Fore coxae of adults 1–1.5 times as broad as middle coxae (Figs. 32, 33). 2
2 (1) First instar nymphs with at least 6 longitudinal rows of dorsal hairs (i.e. pleural hairs present); thoracic segments with two pairs of marginal hairs; ant. segm. II with more than one hair. .... **Phyllaphidina** (p. 23)
– First instar nymphs with 4 longitudinal rows of dorsal hairs (i.e. pleural hairs absent); thoracic segments with one pair of marginal hairs; ant. segm. II with one hair. .... **Callaphidina** (p. 46)

Key to genera of Phyllaphidini

Apterous and alate viviparous females

1 Most body segments with spinal and marginal, slender, finger-shaped processes which are curved to the rear (Fig. 112). .... *Ctenocallis* Klodnitzki (p.61)
– Finger-shaped processes absent or only present on dorsum of a few body segments. 2
2 (1) Fore coxae 1.75–3 times as broad as middle coxae (Fig. 31). On Leguminosae. .... *Therioaphis* Walker (p. 67)
– Fore coxae 1–1.5 times as broad as middle coxae (Figs. 32, 33). Not on Leguminosae. 3
3 (2) Processus terminalis nearly twice as long as ant. segm. III, up to about 9 × VIa. .... *Monaphis* Walker (p. 44)
– Processus terminalis much shorter than 2 × segm. III, shorter than 4 × VIa. 4
4 (3) Cauda broad, rounded, semicircular (Fig. 36). Brown aphids on branches of *Betula*. Antennae dark with 2–3 white transverse bands (Fig. 34). .... *Symydobius* Mordvilko (p. 23)
– Cauda knobbed (Fig. 37) or broadly triangular (Fig. 70). If the aphid is brown and lives on branches of *Betula*, then antennae dark without white transverse bands (Fig. 35). 5
5 (4) Alate viviparous females. 6
– Apterous viviparous females. 18
6 (5) Paired spinal finger-shaped or conical processes present on at least one of the anterior abd. tergites I–III, sometimes fused at bases, often pigmented (Figs. 95–97, 121). 7

– Without such processes on anterior abd. tergites. 8
7 (6) Secondary rhinaria narrow, almost linear (Fig. 114). On *Ulmus*. .... *Tinocallis* Matsumura (p. 62)
– Secondary rhinaria broadly oval, almost circular. On *Quercus*. .... *Tuberculatus* Mordvilko (p. 52)
8 (6) Dorsal wax gland plates present. 9
– Dorsal wax gland plates absent. 11
9 (8) Siphunculi pore-shaped. On *Fagus*. .... *Phyllaphis* Koch (p. 30)
– Siphunculi truncate or stump-shaped. On *Betula* or *Alnus*. 10
10 (9) Hairs on ant. segm. III shorter than IIIbd. Colour yellowish green, sometimes with dark cross bars on abdomen. . *Euceraphis* Walker (p. 27)
– Hairs on ant. segm. III longer than IIIbd. Colour brown or greenish black. .... *Clethrobius* Mordvilko (p. 25)
11 (8) Processus terminalis longer than VIa. 12
– Processus terminalis shorter than VIa. 14
12 (11) Siphunculi densely covered with transverse rows of spinules (Fig. 54). .... *Callipterinella* van der Goot (p. 32)
– Siphunculi without spinules. 13
13 (12) Frons almost straight. Antennae as long as or shorter than body. Spinal hairs on abd. tergites in small paired groups of 2–6, each group situated on a small slightly convex area. .... *Myzocallis* Passerini (p. 48)
– Frons concave. Antennae longer than body. Spinal hairs not in small compact paired groups. .... *Calaphis* Walsh (p. 36)
14 (11) Cauda broadly triangular. Hind femur without black spot near knee. On *Betula*. .... *Betulaphis* Glendenning (p. 40)
– Cauda knobbed. Hind femur with black spot near knee or with distal part dark all over. Not on *Betula*. 15
15 (14) Fore wing with dark anterior edge and dark spots at apices of oblique veins. 16
– Fore wing without dark anterior edge, but with dark spot at base of cubitus 2, sometimes also at bases of cubitus 1 and media. 17
16 (15) Siphuncular base with 3–4 hairs. Abdomen with broad dark dorsal cross bars. Head and pronotum without lateral longitudinal stripes. On *Juglans*. .... *Callaphis* Walker (p. 46)
– Siphunculi without hairs. Abdominal dorsum with 2–4 longitudinal rows of dark spots. Head and pronotum with lateral longitudinal stripes. On *Tilia*. .... *Eucallipterus* Schouteden (p. 66)
17 (15) Processus terminalis less than half as long as VIa. Anterior abd. tergites without protuberances. Middle femur usually with a small dark spot near the knee. On *Juglans*. ... *Chromaphis* Walker (p. 47)
– Processus terminalis at least half as long as VIa. Anterior

abd. tergites with inconspicuous median low protuberances. Middle femur without dark spot near the knee. On *Alnus*. .... *Pterocallis* Passerini (p. 57)

18 (5) Dorsal wax gland plates present. .... *Phyllaphis* Koch (p. 30)

– Dorsal wax gland plates absent. 19

19 (18) Siphunculi densely covered with transverse rows of spinules. .... *Callipterinella* van der Goot (p. 32)

– Siphunculi without spinules. 20

20 (19) Processus terminalis more than tvice as long as VIa. Ant. segm. III with rhinaria. .... *Calaphis* Walsh (p. 36)

– Processus terminalis about as long as VIa or shorter. Ant. segm. III without rhinaria. 21

21 (20) Cauda knobbed. On *Alnus*. .... *Pterocallis* Passerini (p. 57)

– Cauda broadly triangular, conical. On *Betula*. ... *Betulaphis* Glendenning (p. 40)

SUBTRIBE PHYLLAPHIDINA

Fore coxae not noticeably larger than middle coxae. – First instar nymphs with at least six longitudinal rows of dorsal hairs on thorax and abd. segm. I–VI (i.e. pleural hairs present); thoracic segments with two marginal hairs on each side; ant. segm. II with 2 or 3 hairs; border between head and pronotum nearly always distinct.

Most of the Scandinavian species feed on *Betula*, some on *Fagus* or *Alnus*.

Genus *Symydobius* Mordvilko, 1894

Symydobius Mordvilko, 1894: 65.
Type-species: *Aphis oblonga* von Heyden, 1837.
Survey: 415.

Five species in the world, one species in Scandinavia.

58. ***Symydobius oblongus*** (von Heyden, 1837)
Plate 2:2. Figs. 33, 34, 36, 38.

Aphis oblonga von Heyden, 1837: 298.
Aphis fuscipennis Zetterstedt, 1840: 312.
Survey: 415.

Apterous viviparous female. Shining brown, without wax powder. Abdomen with dorsal sclerotized cross bands and marginal sclerites. Antennae brown, with basal parts of segm. IV and V white, sometimes also with basal part of segm. VI white. Siphunculi paler than body, often white. Cauda dark. Eyes dark red or black. Body with numerous short, pointed hairs. Frons almost straight. Ocelli present as in alate individuals. Margins of meso- and metathorax with dark swellings. Antenna a little shorter than body; processus terminalis shorter than VIa; segm. III with 10–26 small, transverse oval secondary rhinaria on basal ½–¾; longest hair of segm. III about 0.5 × IIIbd. Apical segm. of rostrum shorter than 2sht., blunt, with about 14–16 accessory hairs. Siphunculus low, truncate, with transverse rows of fine spinules (Fig. 38). Cauda semicircular (Fig. 36). 2.0–3.5 mm.

Alate viviparous female. Ant. segm. III with 16–28 secondary rhinaria; longest hair on segm. III about as long as IIIbd. Wings well developed or shortened. Otherwise similar to the apterous viviparous female.

Oviparous female. Antenna sometimes without pale parts; segm. III with 5–18 small, roundish or oval secondary rhinaria. Hind legs darker than fore and middle legs. Hind tibiae slightly swollen, with many scent plaques.

Apterous male. Rather dark. Slender. Secondary rhinaria on ant. segm. III: 33–38, IV and V: 0. 1.8–2.0 mm.

Distribution. In Denmark widespread and common; in Sweden common all over the country, north to T. Lpm.; in Norway common in the southern part of the country, north to On; in Finland common and widespread, north to Li. – Widespread in W, C & E Europe, including the British Isles, France, Germany, Hungary, Czechoslovakia, and Poland; in Russia north to Khibiny Mountains; in Asia east to Mongolia.

Biology. The aphids form colonies on the bark of usually 3–4 years old branches and twigs of birch (*Betula verrucosa, B. pubescens*). They are attended by ants. If disturbed, the aphids do not let themselves fall to the ground, but escape by crawling away. The tarsal claws are strongly developed and well adapted for crawling on rough surfaces and anchoring. In Denmark the eggs hatch in April or May. The fundatrices are partly apterous, partly alate. Apterous viviparous females, alate viviparous females with normal wings, and alate viviparous females with short or rudimentary wings occur from May till October. Adult sexuales appear in early October. Copulation and oviposition have been observed in October and November.

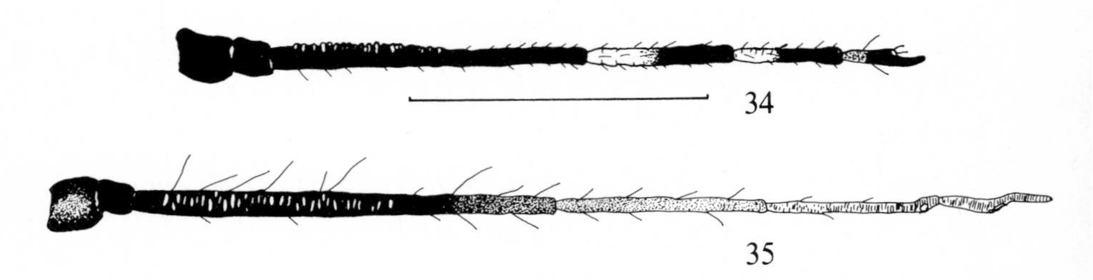

Figs. 34, 35. Antennae of alate viviparous females. – 34: *Symydobius oblongus* (v. Heyd.); 35: *Clethrobius comes* (Wlk.). (Scale 1 mm).

Genus *Clethrobius* Mordvilko, 1928

Clethrobius Mordvilko, 1928: 181, 184.
Type-species: *Callipterus giganteus* Cholodkovsky, 1899.
= *Aphis comes* Walker, 1848.
Survey: 159.

The genus is similar to *Symydobius* with regard to colour, chaetotaxy, and habitat, and similar to *Euceraphis* with regard to shape of cauda, presence of wax glands, and by the fact that all viviparous females are alate.

If the "Survey" (Eastop & Hille Ris Lambers 1976) is to be followed there are two species in the world, both occurring in Scandinavia, viz. *comes* (Walker) living on *Betula* and *giganteus* (Cholodkovsky) living on *Alnus*. They are here, in accordance with Ossiannilsson (1959) and others, regarded as synonyms because they apparently cannot be separated morphologically. In some areas, e.g. in Denmark, most records are from

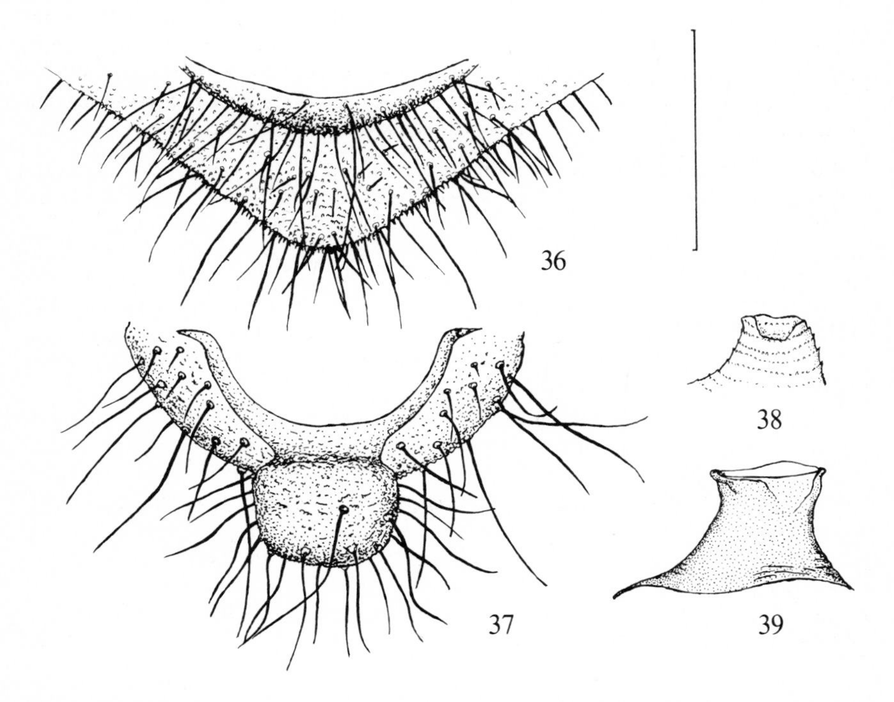

Figs. 36, 37. Cauda and anal plate, dorsal view. – 36: *Symydobius oblongus* (v. Heyd.); 37: *Clethrobius comes* (Wlk.). (Scale 0.2 mm).
Figs. 38, 39. Siphunculi. – 38: *Symydobius oblongus;* 39: *Clethrobius comes*. (Scale 0.2 mm).

Betula, in other areas, e.g. in Great Britain, *Alnus* seems to prevail as the host plant. This may indicate differences between the populations with regard to host preferences. As far as I know it has not yet been tried to transfer the aphids from *Betula* to *Alnus,* or vice versa.

59. ***Clethrobius comes*** (Walker, 1848)
Plate 2: 4. Figs. 35, 37, 39.

Aphis comes Walker, 1848 b: 258.
Callipterus giganteus Cholodkovsky, 1899: 474.
Survey: 159 (as two species).

Alate viviparous female. Brown or greenish black. Antennae, legs, and cauda black, except for bases of femora, which are yellowish or pale brown. Eyes red. Siphunculi as dark as body. Abdomen with more or less extended dorsal and marginal sclerites, larger in fundatrices than in females of later generations, where these sclerites may be replaced by small, more or less fused, scleroites, each carrying one hair. Body with numerous long, pointed hairs. Abdomen with marginal tubercles. Antenna about 0.7 × body; processus terminalis about 0.5 × VIa in fundatrices, about 0.75 × VIa in later generations; segm. III with 30–47 transverse oval rhinaria on basal $^1/_3$–$^2/_3$; hairs on segm. III about 1.2–2.0 × the diameter of the segment below the basal rhinarium (Fig. 35). Apical segm. of rostrum about 0.7 × 2sht., blunt, with many hairs. Siphunculus short, truncate, about 0.5 × cauda, with dilated aperture (Fig. 39). Cauda knobbed, distinctly constricted, with about 14–20 hairs (Fig. 37). 4.3–5.3 mm.

Oviparous female. The only apterous morph . Ocelli indicated. Meso- and metathorax with lateral swellings. Antenna rather short, without secondary rhinaria. Hind tibia strongly swollen for almost whole length, with more than a hundred scent plaques. Cauda semicircular, without constriction.

Alate male. Slender. Antenna almost as long as body; segm. III with 67–73 secondary rhinaria, IV and V without secondary rhinaria. About 3.5 mm.

Distribution. In Denmark widespread and rather common, mostly on *Betula,* less common than *Symydobius oblongus;* in Sweden also rather common and widespread, north to T. Lpm., partly on *Betula,* partly – e.g. in Vg., Dlr., and Bl. – on *Alnus;* in Norway north to MRy; in Finland recorded from the southern part on both hosts – Widespread in Europe, but apparently not very common; it is rare in Great Britain (on *Alnus glutinosa,* rarely on *Betula*) and Germany (not recorded from N Germany); recorded from Poland (incl. the Baltic region) (on *Alnus incana*), the Netherlands (*Betula*), Switzerland (*Alnus*), Italy (*Alnus*), Czechoslovakia, and NW & W Russia; in Asia east to Mongolia and perhaps E Himalayas.

Biology. The hosts are *Betula* (*B. verrucosa, B. pubescens, B. papyracea*) and *Alnus* (*A. glutinosa, A. incana*). In Denmark colonies have been found on 1–4 years old branches and twigs from May till October, but not on the same tree for the whole period (Heie 1972c). All individuals usually leave the branch where they were born, as soon as they are full-grown and have wings. Sometimes the colonies are visited by ants. The aphids drop to the ground when disturbed (in contrary to *Symydobius oblongus*). The species may often be overlooked because the colonies are located in the tree-tops. Sexuales have been observed in October.

Genus *Euceraphis* Walker, 1870

Euceraphis Walker, 1870: 2001.
Type-species: *Aphis betulae* Linné of Walker, 1848b, nec Linné 1758 = *Aphis punctipennis* Zetterstedt, 1828.
Survey: 194.

Large aphids somewhat similar to *Drepanosiphum,* but with shorter processus terminalis and shorter siphunculi. All viviparous females are alate, including the fundatrices. The genus differs from *Clethrobius* in having shorter hairs, those on ant. segm. III being shorter than basal diameter of the segment.

Seven species in the world. In Scandinavia two species which until 1976 were regarded as one species, *punctipennis.* Blackman (1976, 1977) demonstrated that the populations from *Betula pubescens* and *B. verrucosa* have a different numbers of chromosomes, viz. 8 and 10, respectively. Also small morphological differences were found.

They are not visited by ants.

Key to species of *Euceraphis*

Alate viviparous females

1 VIa 0.29–0.43 mm, usually 1.4–1.8 × 2sht., shorter than 1.37 × 2sht. in some fundatrices, only. .... 61. *punctipennis* (Zetterstedt)
– VIa 0.17–0.33 mm, usually 1.0–1.3 × 2 sht., rarely up to 1.37 × 2sht.; VIa only in specimens with body length 3.4 mm or more longer than 0.29 mm, and then usually shorter than 1.32 × 2sht. ... 60. *betulae* (Koch)

60. ***Euceraphis betulae*** (Koch, 1855)
Figs. 40, 44, 45.

Callipterus betulae Koch, 1855: 217.
Survey: 194 (as a synonym of *E. punctipennis* (Zetterstedt)).

Alate viviparous female. Differs from *punctipennis* in the following characters: VIa usually 0.17–0.29 mm, up to 0.33 mm in large specimens with body length 3.4 mm or more, and then shorter than 1.32 × 2sht. Fundatrices and sexuparae sometimes with dark cross bars on all abd. tergites. Basal parts of ant. segm. III and IV dark in specimens from spring and autumn; distal fourth or more of fore tibia dark and rough in such specimens. Siphunculi sometimes black. 3.0–4.2 mm.

Oviparous female. Very much like *punctipennis,* but VIa 0.19–0.27 mm, shorter than 1.12 × 2sht.

Alate male. Very much like the male of *punctipennis.* VIa 0.24–0.32 mm, shorter than 1.36 × 2sht.

Distribution. In Denmark extremely common and widespread; in Sweden common, north to Med.; in Norway known from SFi, probably widespread; in Finland known from the southern part of the country, Ab, N, Ta, Sa, and Ok. – Common and widespread in Europe, south to Switzerland and Hungary, east to Russia; very common in the British Isles, N Germany, and Poland. Introduced and apparently widespread in N America.

Biology. The host is *Betula verrucosa.* Life cycle and biology as in *punctipennis.*

61. ***Euceraphis punctipennis*** (Zetterstedt, 1828)
Plate 2: 3. Figs. 41–43.

Aphis punctipennis Zetterstedt, 1828: 559.
Aphis nigritarsis von Heyden, 1837: 299.
Survey: 194 (partim; *E. betulae* (Koch) is included).

Alate viviparous female. Pale green, somewhat wax powdered, with tufts of wax threads on antennae and legs. Head with dark dorsal spot or longitudinal stripe. Prothorax with two brown spots, meso- and metathorax brown. Abdomen sometimes with short, black cross bars on tergites IV–V or on IV only, in some fundatrices and sexuparae also with paired sclerites on other tergites. Abdomen with roundish marginal tubercles on segm. I–V. Antenna as long as body; processus terminalis shorter than VIa;VIa 0.29–0.43 mm, (1.3–) 1.4–1.7 (–1.8) × 2sht., longer than 1.32 × 2sht. except in some fundatrices; segm. III with 14–26 narrow transverse oval secondary rhinaria almost on line on basal third, which is slightly thicker than the rest of the segment; secondary rhinaria surrounded by short hairs; basal parts of segm. III and IV pale. Apical segm. of rostrum about 0.7 × 2sht. Distal fourth, or less, of fore tibia dark and rough. First tarsal segments usually with 9 hairs. Siphunculus pale or brownish, truncate, about 0.03 × body, shorter than cauda. Cauda pale or slightly brownish, knobbed, constricted, with 12 hairs on distal part. 3.0–4.8 mm.

Oviparous female. Yellowish brown or dark brown (Plate 2: 3). Abdomen with dark cross bars and spots. The posterior part prolonged, conical, functioning as an ovipositor. Antennae dark, without secondary rhinaria; VIa 0.25–0.36 mm, longer than 1.15 × 2sht. Hind tibia slightly thickened, with numerous scent plaques. Siphunculus dark, rough, with well developed flange (in contrary to the condition in viviparous females). Cauda not constricted.

Alate male. Abdomen green with several dark dorsal cross bars. Slender. Antennae dark; VIa 0.30–0.41 mm, longer than 1.35 × 2sht.; secondary rhinaria oval, on segm. III: 65–80, IV: 0–2, V: 0–19.

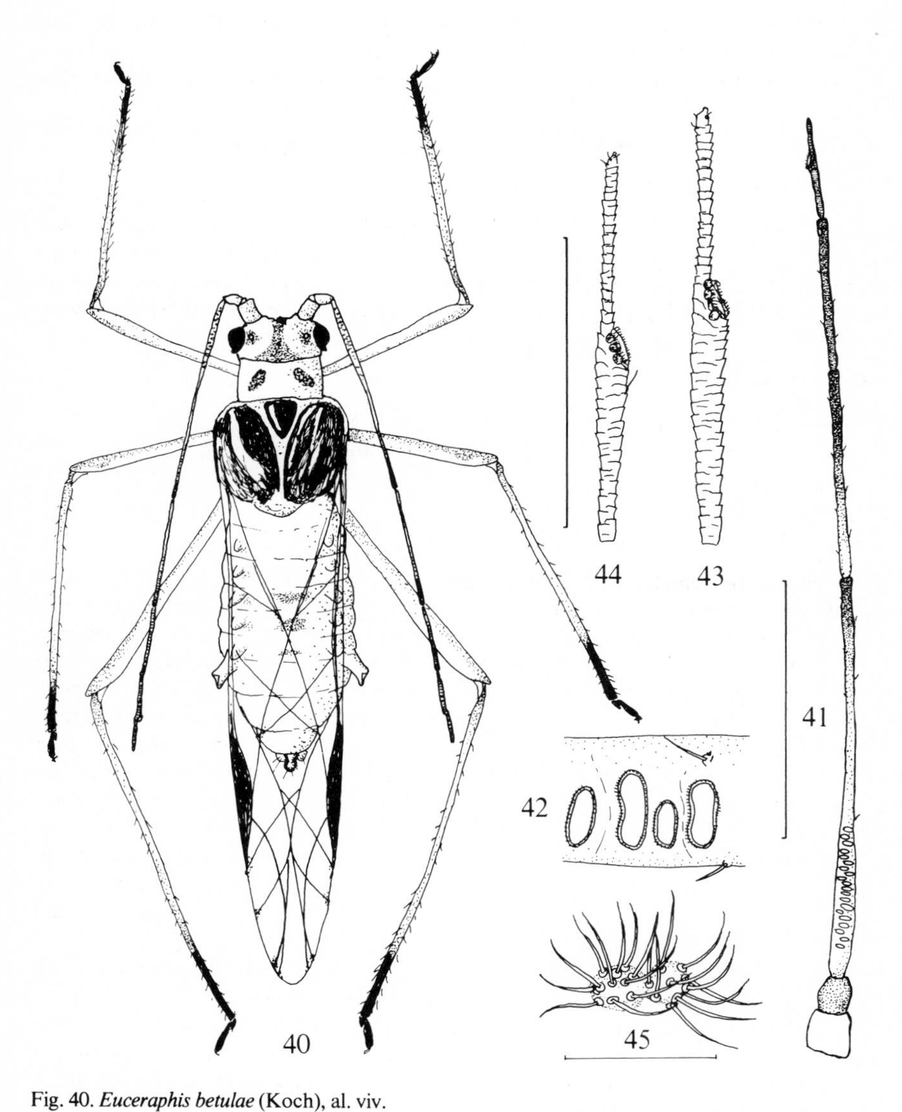

Fig. 40. *Euceraphis betulae* (Koch), al. viv.
Figs. 41–45. *Euceraphis* spp., al. viv. – 41: antenna of *punctipennis* (Zett.); 42: part of ant. segm. III of same showing shapes and hairy rims of secondary rhinaria; 43: ant. segm. VI of same; 44: ant. segm. VI of *betulae* (Koch); 45: rudimentary gonapophyses of same. (Scales 1 mm for 41, 0.5 mm for 43 and 44, 0.1 mm for 45).

Distribution. In Denmark extremely common and known from all districts; very common and widespread in Sweden from Sk. north to T. Lpm.; known from Norway, probably common and widespread; in Finland known from several districts. – Common and widespread in Europe, including the Faroes, Iceland, the British Isles, N Germany, Poland, and Russia, south to Spain; USSR; Greenland; introduced in N America.

Biology. The aphids live on the underside of leaves and on young shoots of *Betula pubescens*. They usually move around and, therefore, seldom form colonies. In Denmark, the eggs usually hatch in early April, in some years as early as in March, in other years first in the beginning of May. Fundatrix normally is adult in late April or in May (Heie 1972c). Sexuales were observed from early September till December. The oviparous female crawls on the bark of branches and twigs, and oviposites on the bark, preferably close to the buds.

Genus *Phyllaphis* Koch, 1857

Phyllaphis Koch, 1857: 248.
Type-species: *Aphis fagi* Linné, 1767.
Survey: 352.

Abdominal segments each with one spinal pair, one pleural pair, and one marginal pair of distinct, honeycomb-like wax gland plates (Fig. 51). Frons convex, without tubercles. Eyes of apterae consisting of rather few (25–60) ommatidia. Secondary rhinaria of alatae surrounded by short hairs. Processus terminalis much shorter than VIa. First tarsal segments normally with 5-5-5 hairs, sometimes with 6 on one or some legs. Siphunculi pore-shaped (Fig. 3). Cauda knobbed. Male alate.

With two species in the world, one species in Scandinavia.

62. ***Phyllaphis fagi*** (Linné, 1767)

Plate 1: 3. Figs. 3, 46–52.

Aphis fagi Linné, 1767: 735. – Survey: 352.

Apterous viviparous female. Pale yellowish green, covered with wax wool. Wax gland plates as described above, pale or dark; pleural and spinal plates more or less fused. Body, antennae, and legs with rather few short hairs. Antenna a little shorter than 0.5 × body, without secondary rhinaria; processus terminalis about 0.11–0.12 × VIa. Rostrum reaches nearly to 2nd coxae; apical segm. a little shorter than 2 sht. Anal plate rounded, very slightly emarginate. 2–3 mm.

Alate viviparous female. Abdomen greenish, with wax pores placed on dark marginal and dorsal sclerites; the latter sometimes fused, forming cross bars; covered with wax. Head and thorax dark, with wax pores. Antenna about 0.75 × body;

processus terminalis 0.10–0.13 × VIa; segm. III with 4–9 transverse oval secondary rhinaria on basal half or mid third. Cauda dark, distal part globular. Anal plate emarginate.

Oviparous female. Pale yellow or yellowish green with dark spots. Antenna a little

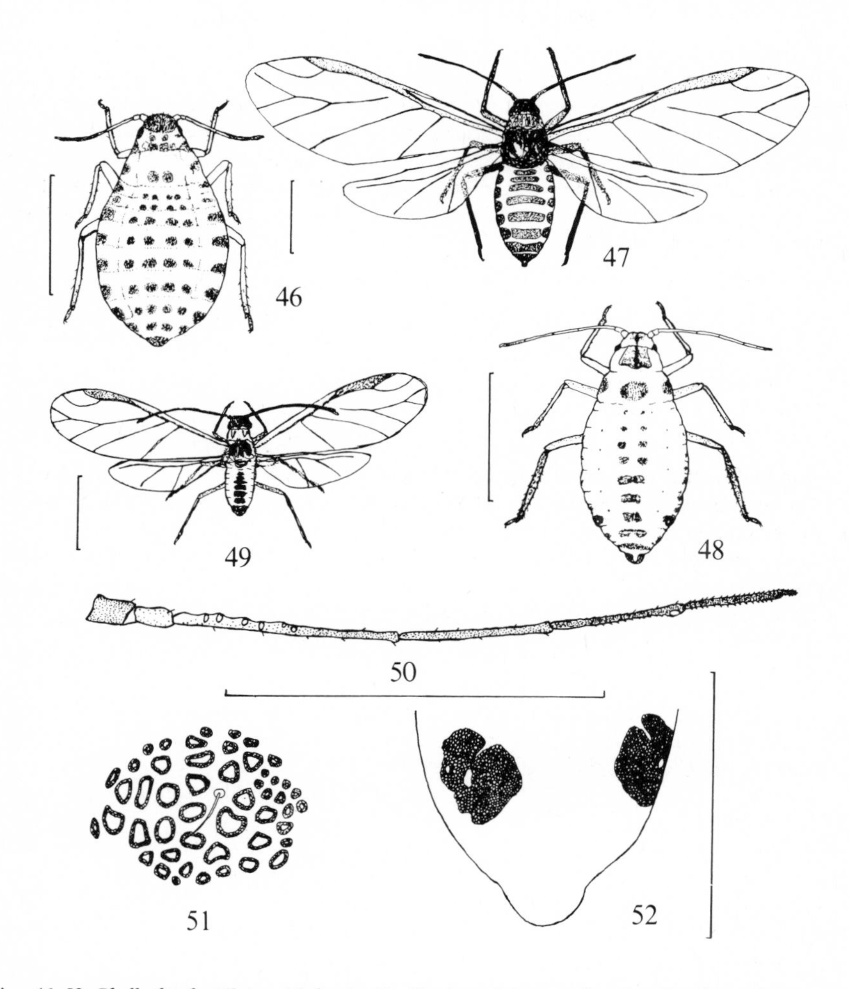

Figs. 46–52. *Phyllaphis fagi* (L.). – 46: fundatrix; 47: alate viviparous female; 48: oviparous female; 49: alate male; 50: antenna of alate viviparous female; 51: dorsal wax gland plate; 52: position of ventral wax gland plates in the oviparous female. (Scales 1 mm for 46–50 and 52). (46–49 after Schmutterer, redrawn).

longer than 0.5 × body, without secondary rhinaria. Hind tibiae only slightly thickened, with 40–80 scent plaques of irregular shapes.

Alate male. Ant. segm. III with 62–89 secondary rhinaria, placed along the whole segment, IV: 35–52, V: 20–33, VI: 8–12. Very much like the alate viviparous female, but smaller, about 1.8 mm.

Distribution. In Denmark extremely common and widespread; in Sweden common in the south, north to Vstm. and Upl. (see map in Heie, 1980: Fig. 53; Sm. shall be added (Danielsson 1974)); in Norway north to SFy and SFi; in Finland found in Helsinki Botanical Garden (N). – Nearly all over the world, where *Fagus* occurs. In Europe south to Portugal, Spain, and Yugoslavia, east to the USSR, in Great Britain common more or less over the whole range of the host, common in N Germany, Poland, and NW & W Russia. N Asia, the Caucasus region, Turkey, Japan, Australia and New Zealand. Widespread in USA and Canada.

Biology. The wax-covered colonies occur on the undersides of leaves of *Fagus silvatica.* Young leaves are preferred and are often bent along the midrib towards the underside. Even the seed leaves (cotyledones) of small plants can be infested. The species is not visited by ants.

Genus *Callipterinella* van der Goot, 1913

Callipterinella van der Goot, 1913: 118.
Type-species: *Aphis betularia* Kaltenbach, 1843
= *Aphis tuberculata* von Heyden, 1837.
Survey: 120.

Dorsal hairs long, strong, pointed or blunt. Antennae shorter than body; processus terminalis longer than base of ultimate segment; segm. III with circular secondary rhinaria in alate viviparous females, in some species also in apterae. Radial sector indistinct or absent; other wing veins dark-bordered. Siphunculi low, truncate, with numerous transverse rows of spinules; siphuncular apertures extended. Cauda in viviparae constricted, knobbed. Anal plate slightly emarginate.

Three species on birch (*Betula*) in the world. They all occur in Scandinavia. They are generally visited by ants.

Key to species of *Callipterinella*

Apterous viviparous females

1 Antenna 5-segmented, without secondary rhinaria. Processus terminalis shorter than twice the length of the basal part of the ultimate ant. segm. .... 64. *minutissima* (Stroyan)

– Antenna 6-segmented, with secondary rhinaria. Processus terminalis about twice as long as basal part of ultimate ant.

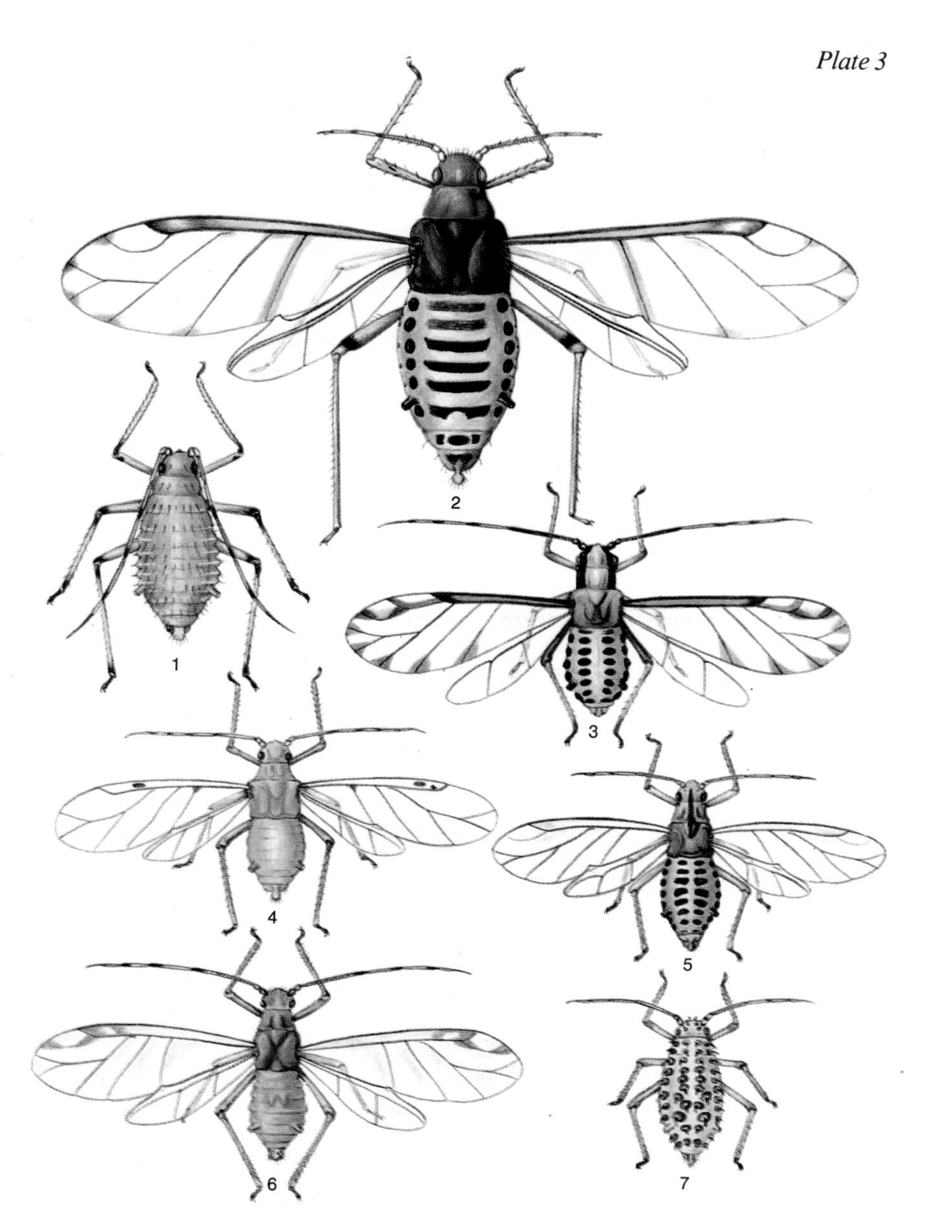

1. *Calaphis flava* Mordv., apt. viv. – 2. *Calaphis juglandis* (Goeze), al. viv. – 3. *Eucallipterus tiliae* (L.), al. viv. – 4. *Myzocallis coryli* (Goeze), al. viv. – 5. *M. myricae* (Kalt.), al. viv. – 6. *Tuberculatus annulatus* (Hartig), al. viv. – 7. *Therioaphis ononidis* (Kalt.), apt. viv. (10 ×).

Plate 4

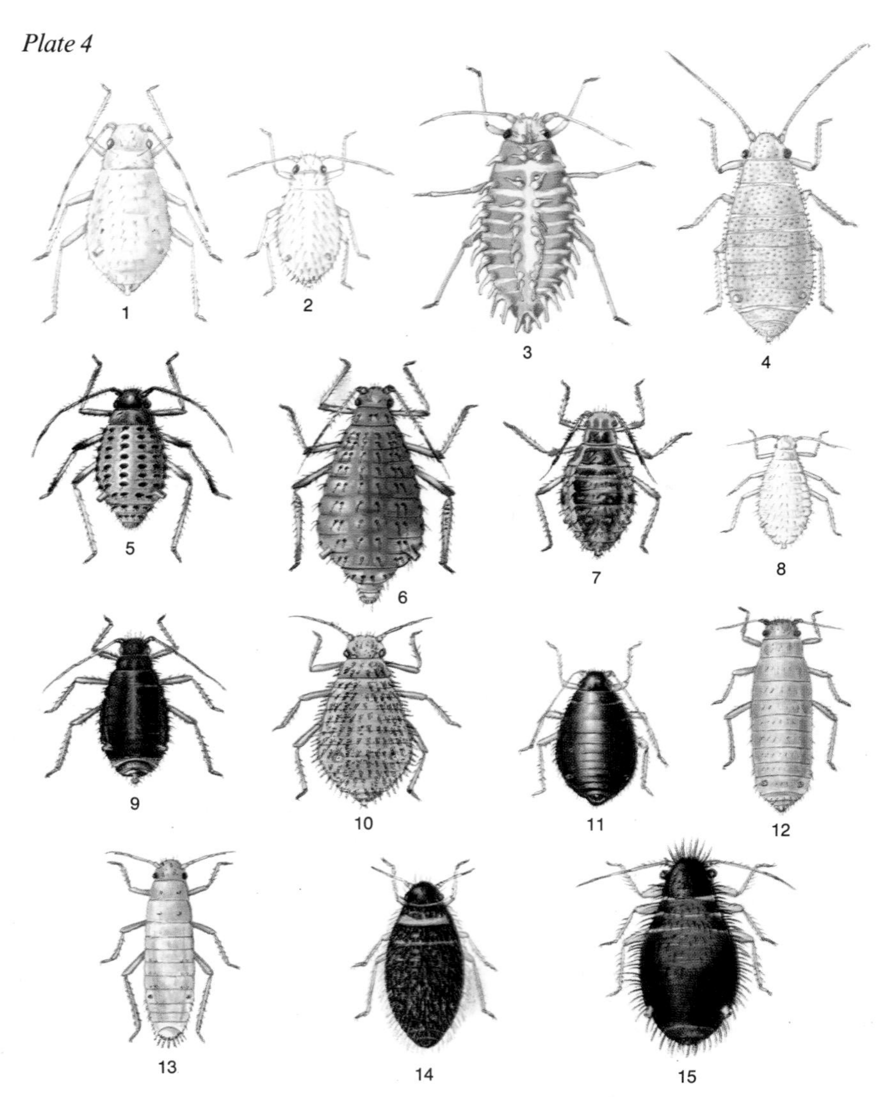

1. *Pterocallis alni* (DeGeer), apt. viv. – 2. *P. albidus* Börn., apt. viv. – 3. *Ctenocallis setosus* (Kalt.), apt. viv. – 4. *Subsaltusaphis flava* (H.R.L.), apt. viv. – 5–6. *Periphyllus testudinaceus* (Fern.), apt. viv. from June (5) and ovip. female from October (6). – 7. *Chaitophorus leucomelas* Koch, apt. viv. – 8. *C. horii* ssp. *beuthani* (Börn.), apt. viv. – 9. *C. tremulae* Koch, apt. viv. – 10. *Sipha glyceriae* (Kalt.), apt. viv. – 11. *S. (Rungsia) maydis* Pass., apt. viv. – 12. *Laingia psammae* Theob., apt. viv. – 13. *Atheroides serrulatus* Hal., apt. viv. – 14. *Chaetosiphella berlesei* (d. Gu.), apt. viv. – 15. *Caricosipha paniculatae* Börn., apt. viv. (1–4: 15 ×, the rest 10 ×).

segm., or longer 2

2 (1) Greenish with blackish brown dorsal cross bars, never with large, black dorsal spot on posterior part of abdomen. .... 63. *calliptera* (Hartig)

– Yellow with red dorsal band, brown spots, and large, black dorsal spot on posterior part of abdomen. .... 65. *tuberculata* (von Heyden)

Alate viviparous females

1 Abdomen with dark dorsal cross bars. .... 63. *calliptera* (Hartig)

– Abdomen without dark dorsal cross bars. 2

2 (1) Ant. segm. III with 12–17 rhinaria. Processus terminalis longer than twice the length of basal part of ultimate ant. segm. .... 65. *tuberculata* (von Heyden)

– Ant. segm. III with 5–8 rhinaria. Processus terminalis shorter than twice the length of basal part of ultimate ant. segm. 64. *minutissima* (Stroyan)

63. ***Callipterinella calliptera*** (Hartig, 1841)

Figs. 53–55, 61.

Aphis calliptera Hartig,1841: 369. – Survey: 120.

Apterous viviparous female. Green with blackish brown dorsal cross bars. Marginal sclerites present, each with one hair. Antennae pale with black apices of segments. Siphunculi dark. Legs brownish, hind legs darker than fore and middle legs. Eyes red. Body and hind tibiae with strong and pointed hairs as long as apical segm. of rostrum. Antenna a little shorter than body; processus terminalis about 2 × VIa; segm. III with 2–5 rhinaria. Siphunculi and abdominal cross bars with densely placed rows of spinules (Fig. 54). 1.6–2.5 mm.

Alate viviparous female. Abdomen with dark dorsal cross bars. Hairs shorter than in apterous viviparous female. Rather large marginal tubercles present on abd. segm. II–V.

Oviparous female. Similar to the apterous viviparous female. Antenna shorter, only a little longer than 0.5 × body; segm. III with 2–4 rhinaria. Hind tibiae swollen, with scent plaques. Cauda not constricted.

Alate male. Rather small and dark. Secondary rhinaria on ant. segm. III: 9–15, IV: 0–1.

Distribution. In Denmark rather common, known from EJ, NWJ, NEJ, and NEZ; in Sweden from Sk. north to Nb.; in Norway found in SFi; in Finland in Sa. – In Europe and Asia east to Mongolia, south to Portugal, Spain, and the Caucasus region; in Great Britain local, but widely distributed; in N Germany common; also known from Poland and NW & W Russia. Introduced in N America.

Biology. The species lives on *Betula verrucosa* and *B. pubescens,* in Denmark preferably on the former species. It seems to depend on attention of ants, and needs also shelter, as found between leaves or parts of leaves spun together by other insects, e.g. larvae of moths (Heie 1972c). Both apterae and alatae are found throughout the summer months. The individuals apparently tend to aggregate, but the colonies are rarely large. Adult sexuales are in Denmark observed in October.

64. ***Callipterinella minutissima*** (Stroyan, 1953)
Figs. 58–60.

Calaphis minutissima Stroyan, 1953: 13. – Survey: 120.

Apterous viviparous female. Greenish, more yellowish green along margins and posteriorly. Vertex with a brown sclerite. A white tuft of wax present below cauda. Abd. tergite VIII with dark cross bar; other tergites with very small scleroites carrying long, fine, pointed hairs. Dorsal hairs on abd. segm. III 3–4 × IIIbd., those on segm. VIII much longer. Antenna 5-segmented, 0.25–0.33 × body; processus terminalis 1.2–1.8 × Va; secondary rhinaria absent; longest hair on segm. III about as long as IIIbd. Siphunculus shorter than wide, with spinules. 0.9–1.4 mm.

Alate viviparous female. Abdomen greenish. Apices of antennae and tibiae, and whole tarsi, dark. Abdomen without marginal tubercles. Antenna 6-segmented, 0.4–0.8 × body; processus terminalis 1.3–1.5 × VIa; segm. III with 5–8 rhinaria forming a row;

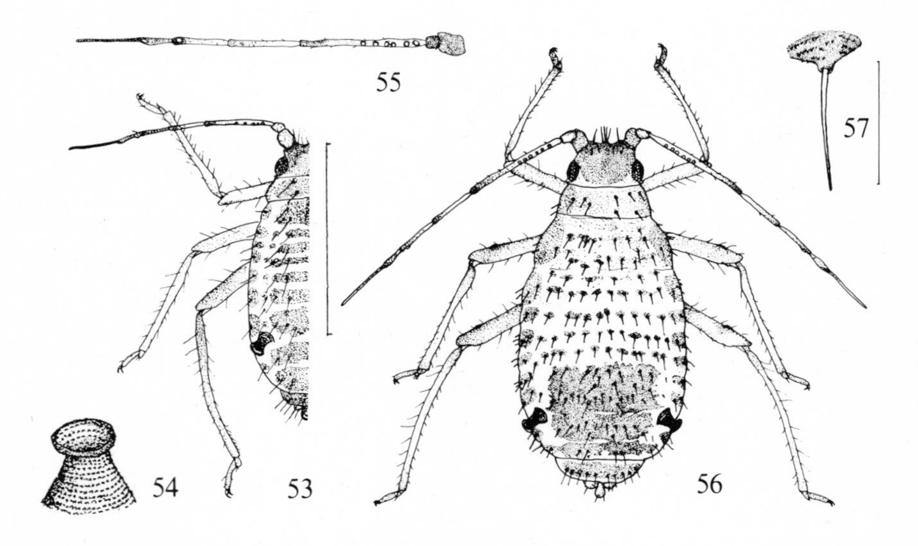

Figs. 53–55, *Callipterinella calliptera* (Hartig). – 53: apt. viv.; 54: siphunculus of apt. viv.; 55: antenna of al. viv.
Figs. 56, 57: *C. tuberculata* (v. Heyd.). – 56: apt. viv.; 57: dorsal hair on scleroite from anterior part of abdomen of apt. viv. (Scales 1 mm for 53 and 56, 0.1 mm for 57).

longest hair on segm. III about 0.8 × IIIbd. Larger than apterous viviparous female, 1.5–1.9 mm.

Oviparous female. Dark green or reddish brown. Abdomen with more or less fused scleroites (Fig. 59). Marginal tubercles may be present on some abdominal segments. Antenna 5- or 6-segmented, without secondary rhinaria. Hind tibia swollen, entire surface with many scent plaques which are often fused two by two. Larger than apterous viviparous female, 1.4–1.9 mm.

Alate male. Dark green with brown head and thorax. Abdomen with marginal tubercles. Antenna 6-segmented, about 0.6–0.7 × body; segm. III with 7–11 rather large rhinaria placed in a row, IV and V without secondary rhinaria. 1.3–1.7 mm.

Distribution. In Denmark found on a single tree in NWJ (Skive), in Sweden found in Sk. by Danielsson; not in Norway and Finland. – In W, C & S Europe, very rare; in Great Britain in Surrey, Hertford, and Cambridge; in Germany only in Berlin; also records from Bohemia in Czechoslovakia and from Portugal (Villa Nova de Paiva near Porto).

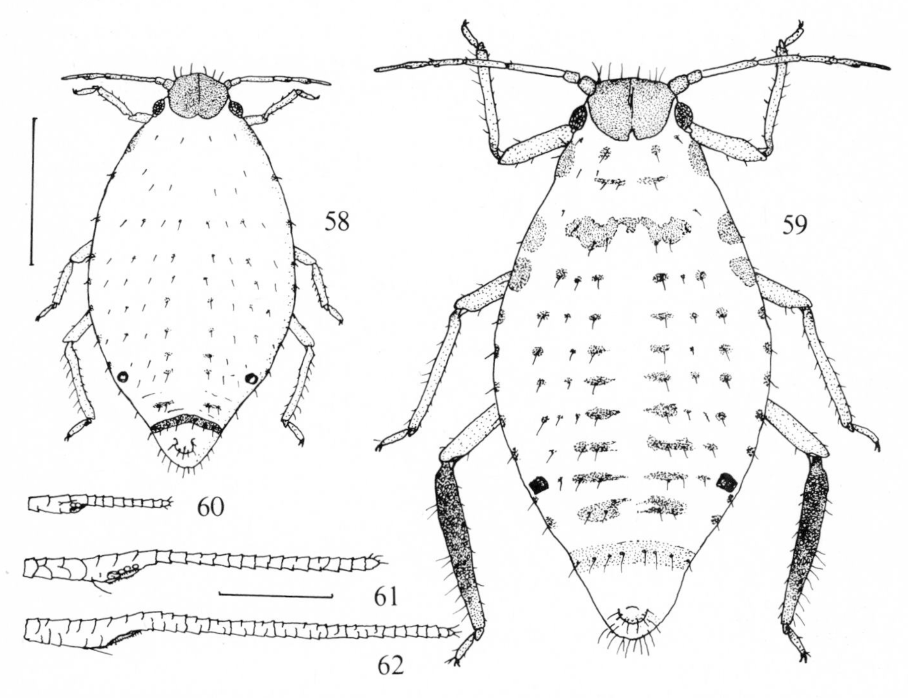

Figs. 58–62. *Callipterinella* spp., apt. viv. – 58: *minutissima* (Stroyan), apterous viviparous female; 59: same, oviparous female; 60–62: ultimate ant. segm. of *minutissima* (60), *calliptera* (Hartig) (61), and *tuberculata* (v. Heyd.) (62). (Scales 0.5 mm for 58 and 59, 0.1 mm for 60–62).

Biology. The species has been found on a very few birch trees (*Betula pubescens, B. verrucosa*), in Denmark on a single tree (*B. pubescens*) in my garden, mainly in autumn, rarely, or not, during the summer months. In 1977 the sexuales were particularly abundant in the beginning of November.

65. ***Callipterinella tuberculata*** (von Heyden, 1837)
Plate 1: 6. Figs. 56, 57, 62.

Aphis tuberculata von Heyden, 1837: 296. – Survey: 120.

Apterous viviparous female. Yellow with brown head; pro- and mesothorax with dorsal sclerites; anterior part of abdomen red; posterior part of abdomen (tergites IV–VI) with large, black dorsal spot; marginal sclerites present, each with 3–7 hairs. Antennae, legs, and siphunculi dark except for basal half of ant. segm. III. Dorsal hairs on anterior part of abdomen long, very strong, and blunt (Fig. 57), placed on small, brown scleroites, shorter than apical segm. of rostrum; dorsal hairs on posterior part of abdomen pointed and nearly as long as or as long as apical segm. of rostrum. Antenna a little shorter than body; processus terminalis about 2.5 × VIa; segm. III with 4–8 rhinaria on basal half. Siphunculi and dorsal sclerites with densely placed rows of spinules. 1.7–2.2 mm.

Alate viviparous female. Abdomen with small, brown dorsal spots, but without cross bars and larger spots. Marginal tubercles as in *calliptera.* Dorsal hairs shorter than in apterous viviparous female. Ant. segm. III with 12–17 rhinaria along the entire segment.

Oviparous female. Rather similar to apterous viviparous female.

Apterous male. Very much like the alate viviparous female, but without wings. Secondary rhinaria usually present only on ant. segm. III.

Distribution. In Denmark rather common in the Copenhagen area (NEZ), but not found in other parts of the country; in Sweden rather common and widespread, north to Upl.; rather common in Norway north to TR; in Finland common in the south: N, Ta, Sa, Tb. – In Europe and Asia, east to Siberia, south to Yugoslavia and Kazachstan; in Great Britain local and not very common; common in the Netherlands, N Germany, and Poland; also recorded from NW & W Russia.

Biology. The host is *Betula verrucosa.*

Genus *Calaphis* Walsh, 1863

Calaphis Walsh, 1863: 301.
Type-species: *Calaphis betulella* Walsh, 1863.
Kallistaphis Kirkaldy, 1905: 417.
Type-species: *Aphis betulicola* Kaltenbach, 1843.
Survey: 119.

Delicately built aphids with rather long, thin legs. Lateral frontal tubercles well developed. Body hairs long and capitate in nymphs and apterous adults, short and blunt in alate individuals. Antenna longer than body; processus terminalis longer than VIa; secondary rhinaria circular or subcircular, present on ant. segm. III in all morphs. Radial sector indistinct or absent. Siphunculi low, truncate, with extended apertures. Cauda slightly constricted. Anal plate bilobed in viviparae.

There are 18 species in the world, two species in Scandinavia. The hosts are amentaceous trees (Betulaceae, Myricaceae, Fagaceae). They are not visited by ants.

Key to species of *Calaphis*

Apterous viviparous females

1 Secondary rhinaria placed on basal half of ant. segm. III; distance between base of segment and basal rhinarium only 5–11% of length of segment (Fig. 66). Apical segm. of rostrum of about the same length as 2sht. .... 67. *flava* Mordvilko

– Secondary rhinaria placed on middle of ant. segm. III; distance between base of segment and basal rhinarium 21–35% of length of segment (Fig. 65). Apical segm. of rostrum longer than 2sht. .... 66. *betulicola* (Kaltenbach)

Alate viviparous females

1 Siphunculi usually pale all over, rarely with dark apices. Antenna shorter than 1.4 × body. Ant. segm. III with rhinaria on basal half or slightly more; distance between base of segment and basal rhinarium 5–11% of length of segment (Fig. 64). .... 67. *flava* Mordvilko

– Apical half of siphunculus black. Antenna about 1.4 × body. Ant. segm. III with rhinaria on middle section, which is more or less darkened; distance between base of segment and basal rhinarium 13–24% of length of segment (Fig. 63). .... 66. *betulicola* (Kaltenbach)

66. ***Calaphis betulicola*** (Kaltenbach, 1843)
Figs. 63, 65, 68.

Aphis betulicola Kaltenbach, 1843: 44. – Survey: 119.

Apterous viviparous female. Pale green. Apices of ant. segments dark. Knees and apices of legs dark. Apical half of siphunculus usually dark. Dorsal body hairs capitate, placed on wart-like elevations. Antenna longer than 1.3 × body; segm. III with 5–14 rhinaria on middle third (Fig. 65). Apical segm. of rostrum 0.15–0.17 mm, distinctly longer than 2sht. 2.0–2.3 mm.

Alate viviparous female. Ant. segm. III with 10–18 rhinaria placed on the darkened middle area; antenna distal to this area also brown (Fig. 63). Wing veins darker than in *flava*. Apical half of siphunculus black all over.

Distribution. Apparently rare in Denmark, three records from Jutland (EJ, NEJ); in Sweden widespread, from the southern part of the country north to Vb.; in Norway caught in trap in AK; in Finland known from N, Ok and ObN. – Not very common in Europe and Asia, east to Siberia, south to Spain; in Great Britain widespread and probably not uncommon, but often overlooked; recorded from Germany, including N Germany, Poland, Czechoslovakia, Hungary, and NW Russia. According to Quednau (1966) also in California.

Biology. The aphids live scattered on the undersides of leaves of *Betula verrucosa* and *B. pubescens*. According to Stroyan (1977) they almost only occur on seedlings and small trees less than 1 m high. This may explain why this species is found less frequently than *flava*.

Note. Until 1957 the name *betulicola* was used for both *betulicola* Kalt. and *flava* Mordv. (= *basalis* Stroyan) in Europe (see notes to *flava*).

67. ***Calaphis flava*** Mordvilko, 1928
Plate 3: 1. Figs. 64, 66, 67, 69.

Calaphis flava Mordvilko, 1928d: 184.
Kallistaphis basalis Stroyan, 1957: 343.
Survey: 119.

Apterous viviparous female. Pale green or yellowish. Antennae pale with dark apices of segments. Legs pale with knees, apices of tibiae, and tarsi, dark. Siphunculi pale (Fig. 69). Dorsal body hairs long, capitate, placed on wart-like elevations. Antenna a little longer than body, up to 1.33 × body; processus terminalis more than twice as long as VIa; segm. III with rhinaria on basal half (Fig. 66). Apical segm. of rostrum 0.13–0.15 mm, of about the same length as 2sht. 1.9–2.7 mm.

Alate viviparous female. Ant. segm. III with 6–14 rhinaria on basal $^{3}/_{5}$ (Fig. 64). Wing veins strong, black-bordered, except for the reduced radial sector (Fig. 67). Siphunculi pale, rarely with dark apices.

Oviparous female. Much like the apterous viviparous female, but often with dark dorsal markings and dark siphunculi. Ant. segm. III with 3–7 rhinaria. Hind tibia somewhat thickened, with numerous scent plaques. Subsiphuncular wax gland plates absent.

Alate male. Darker and more slender than alate viviparous female. Abdomen with dark, broad, more or less fused cross bars. Siphunculi sometimes dark. Ant. segm. III with 11–21 rhinaria on basal $^{4}/_{5}$. Radial sector sometimes just as distinct as the other wing veins.

Distribution. In Denmark very common and widespread; in Sweden also very common and widespread, from Sk. north to Lu.Lpm.; common in the southern part of

Norway; in Finland known from N, Ta, Sa, and Oa. – Common in large parts of Europe and Asia, east to Siberia and C Asia, south to Portugal and Hungary; widespread and locally abundant in Great Britain; common in N Germany and Poland, and also known from NW and W Russia. Introduced in Australia, New Zealand and N America.

Biology. The hosts are *Betula pubescens* and *B. verrucosa,* in Denmark preferably the former. *B. pubescens* may be strongly infested, especially in early summer. *B. verrucosa* is usually infested by only a few individuals (Heie, 1972c). Young leaves are preferred, so that birch-hedges which have been cut may be heavily infested on the new summer-shoots. At this time the populations on other birches usually have dropped following the fore summer maximum. Alate females occur in all viviparous generations, also among fundatrices. In Denmark sexuales have been seen from September till November. In September apterous viviparous females which are similar to oviparous females may occur, with a few scent plaques on the hind tibiae. These are, however, not thickened as is the case of oviparous females.

Note. Until recently most European authors did not separate this species from *betulicola* (Kalt.). Stroyan (1957) discovered that two species were mixed together under that name and named the species with secondary rhinaria on the basal part of ant. segm. III as *basalis.* It was, however, already described as *flava* by Mordvilko (1928) from Europe and as *granovskyi* by Palmer (1952) from N America, and is much more common than *betulicola.*

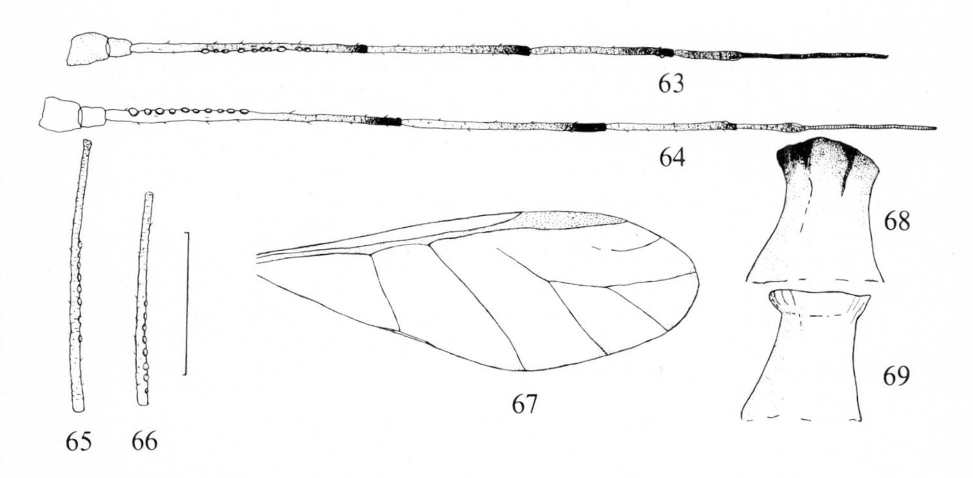

Figs. 63–69. *Calaphis* spp. – 63: antenna of al. viv. of *betulicola* (Kalt.); 64: antenna of al. viv. of *flava* Mordv.; 65: ant. segm. III of apt. viv. of *betulicola;* 66: ant. segm. III of apt. viv. of *flava;* 67: fore wing of *flava;* 68: siphunculus of apt. viv. of *betulicola;* 69: siphunculus of apt. viv. of *flava.* (Scale 0.5 mm for 65 and 66). (65 and 66 after Stroyan, redrawn, others orig.).

Genus *Betulaphis* Glendenning, 1926

Betulaphis Glendenning, 1926: 96.
Type-species: *Betulaphis occidentalis* Glendenning, 1926
= *Aphis quadrituberculata* Kaltenbach, 1843.
Survey: 105.

Small, rather flat, oval aphids. Antennae shorter than body; processus terminalis about as long as VIa; apterous females without secondary rhinaria; alate females with transverse, oval or subcircular, secondary rhinaria on ant. segm. III; males with secondary rhinaria on segm. III–V. Rostrum rather short and thick, reaching just past fore coxae. Siphunculi low, truncate. Cauda short, conical. Anal plate bilobed in viviparous females. Males apterous. Fundatrices partly alate, partly apterous.

Eight species in the world, three species in Scandinavia. They feed on *Betula* and are usually not visited by ants.

Key to species of *Betulaphis*

(based on Hille Ris Lambers (1952) and Shaposhnikov (1964) with regard to characters of *B. pelei*).

Apterous viviparous females

1 Cuticle smooth, not wrinkled or granulate. Marginal hairs on abd. segm. III 1.0–1.5 × IIIbd. On *Betula nana*. ...... 69. *pelei* Hille Ris Lambers

– Cuticle wrinkled, granulate, or smooth; if smooth, then marginal hairs on abd. segm. III 1.5–2.7 × IIIbd. On *Betula verrucosa* or *B. pubescens*. 2

2 (1) White, yellow or pale green. Abdomen with long capitate hairs on all segments, at least along the margins, usually also on dorsum. Cuticle not distinctly wrinkled or granulate. On undersides of leaves of *Betula pubescens* and *B. verrucosa*. 70. *quadrituberculata* (Kaltenbach)

– Green. Abdomen with long hairs only on posterior segments, on dorsum usually on segm. V–VIII, or only VIII, sometimes on segm. IV–VIII, along margins on segm. IV–VII. Cuticle more or less wrinkled or granulate, particularly along margins of body. On upper- and undersides of leaves of *Betula verrucosa*. .... 68. *brevipilosa* Börner

Alate viviparous females

1 Abdomen with very short, almost invisible hairs, except for along margins of posterior segments; usually with dark dor-

sal spot and dark marginal sclerites. Ant. segm. III with 12–19 rhinaria. .... 68. *brevipilosa* Börner

– Abdomen with rather long hairs, with or without dark dorsal spot. Ant. segm. III with 7–16 rhinaria. 2

2 (1) Dorsal hairs rather thick, at least some of them blunt or capitate. All secondary rhinaria similar, transverse oval, arranged in a row. .... 70. *quadrituberculata* (Kaltenbach)

– Dorsal hairs fine, pointed. Secondary rhinaria transverse oval, with constriction, or divided into two rounded rhinaria which form two rows. .... 69. *pelei* Hille Ris Lambers

68. ***Betulaphis brevipilosa*** Börner, 1940
Plate 1: 8. Figs. 71, 74, 76.

Betulaphis brevipilosa Börner, 1940: 2. – Survey: 105.

Apterous viviparous female. Usually grass-green, sometimes paler. Tarsi and apices of antennae dark. Cuticle usually distinctly wrinkled. Abdomen with shorter hairs than in *quadrituberculata,* except for the marginal hairs on segm. (IV–) V–VIII (or only on VIII); marginal hairs on abd. segm. III 0.2–1.0 × IIIbd.; spinal hairs on abd. segm. VII 0.4–0.7 × IIIbd., or absent. Frons with 6 capitate hairs, 1.3–1.8 × IIIbd., placed on four tuber-

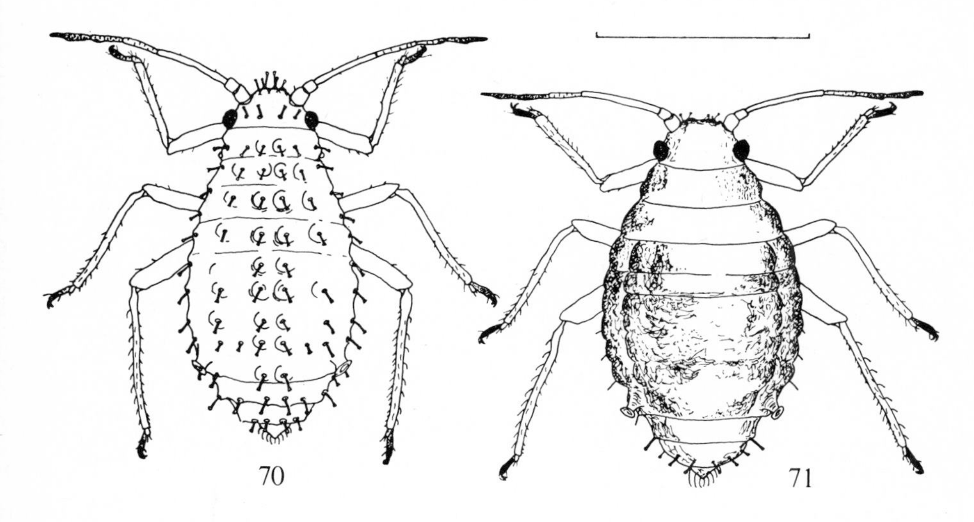

Figs. 70, 71. Apterous viviparous females of 70: *Betulaphis quadrituberculata* (Kalt.) and 71: *B. brevipilosa* Börner. (Scale 1 mm).

cles; two of these are rather broad, each with two hairs. Apical segm. of rostrum shorter than 2sht. 1.5–2.0 mm.

Alate viviparous female. Abdomen usually grass-green, sometimes yellowish, normally with dark dorsal spot on tergites III–V (–VI) and dark marginal sclerites. Siphunculi sometimes dark. Dorsal hairs fine, short, on abd. tergite VI up to 0.6 × IIIbd., or apparently absent from tergites I–VII. Antenna darker than in the apterous viviparous female; segm. III with 12–19 secondary rhinaria (Fig. 74).

Oviparous female. Similar to the apterous viviparous female, but with brown or shining black dorsal cross bars, which may be fused so that the dorsum becomes shining black all over. Cuticle distinctly wrinkled. Posterior part of abdomen prolonged, conical. Hind tibiae not thickened or only slightly so, with rather few scent plaques.

Distribution. In Denmark common and widespread; in Sweden recorded from Sk. in the south to Jmt. in the north; in Norway known from Bø and SFi; in Finland widespread, north to Li. – C Europe; common in N Germany; not recorded from the British Isles; in Poland recorded from the Baltic region. Forbes, Frazer & Chan (1974) recorded *brevipilosa* from British Columbia.

Biology. The species lives on *Betula verrucosa*. In summer dense colonies can be found on the leaves, especially on their uppersides. The life cycle is as in *quadrituberculata*.

Note. Börner's description is very short. Only two characters are mentioned, viz. the very short spinal and pleural hairs and the colour ("mit oder ohne Grünfleckung"). My material is similar to *Betulaphis helvetica* Hille Ris Lambers, 1947 (originally described as a subspecies of *B. quadrituberculata*), of which I have seen part of the type material in Hille Ris Lambers' collection, and which I suppose to be a synonym of *brevipilosa* (see also Quednau 1954). Unfortunately, Börner's types have not been available, so, if there are two species, then the name of the species described above shall be changed to *helvetica*.

69. ***Betulaphis pelei*** Hille Ris Lambers, 1952
Figs. 72, 75.

Betulaphis pelei Hille Ris Lambers, 1952: 23. – Survey: 106.

Apterous viviparous female. Cuticle smooth, sclerotic, and either quite pale, or smoky on meso- and metathorax and brownish on the middle abd. tergites. Abd. segments with very short, rather thick, blunt dorsal hairs and somewhat longer marginal hairs; dorsal hairs on anterior tergites 0.5–0.6 × IIIbd., spinal hairs on tergite VI about 1.0 × IIIbd.; the six hairs on tergite VIII 2.0–2.2 × IIIbd.; marginal hairs on tergites III–IV (–V) 1.0–1.5 × IIIbd.; on VI–VII about 2 × IIIbd. Frons with 3 pairs of capitate hairs, about 2.2 × IIIbd., each hair placed on a conical tubercle (Fig. 75). Antenna about 0.5 × body; processus terminalis a little shorter than VIa. 1.4–1.7 mm.

Alate viviparous female. Abdomen with brownish marginal sclerites, sometimes with transverse rows of intersegmental sclerites between tergites III and IV and between V and VI. Dorsal hairs rather long, fine, pointed. Antenna about 0.7 × body, rather dark brown; segm. III with 7–12 rhinaria on basal two thirds.

Distribution. In Sweden found in T. Lpm. (Abisko); in Norway known from Bv, On, HOi, and MRy; in Finland known from Ok and Le (Heikinheimo leg., Heie det.); not in Denmark. – Greenland (Sarqaq), N Russia (Khibiny Mountains).

Biology. The species lives on *Betula nana*.

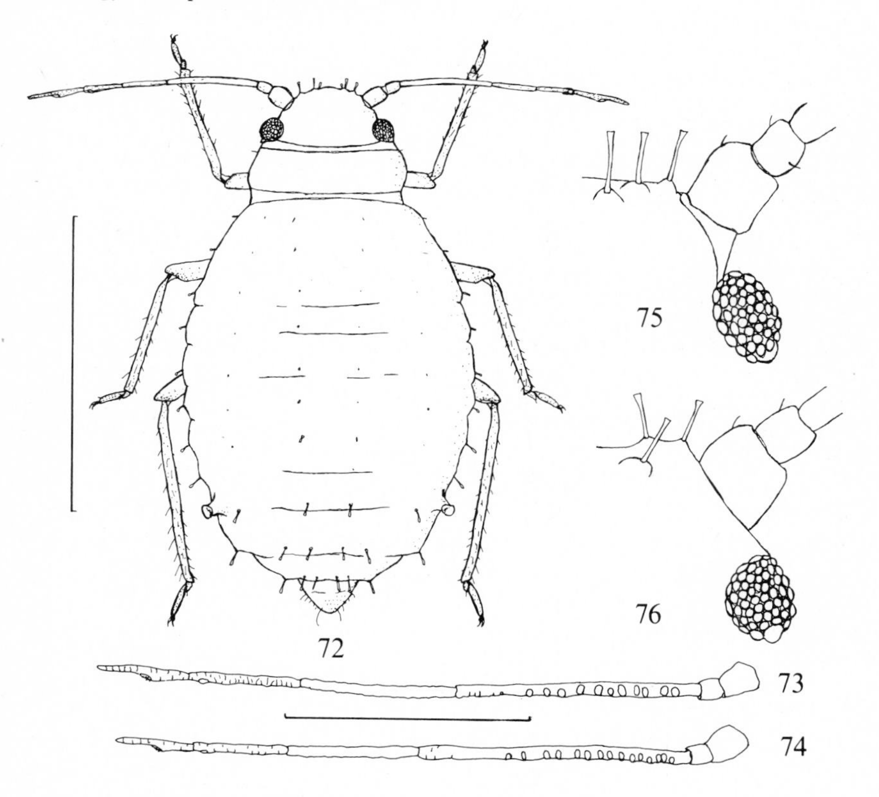

Fig. 72. *Betulaphis pelei* (H. R. L., apt. viv. (specimen from Finland, Le: Enontekiö Kaaresuvanto, on *Betula nana* in July). (Scale 1 mm).
Figs. 73, 74. Antennae of al. viv. of 73: *B. quadrituberculata* (Kalt.) and 74: *B. brevipilosa* Börner.
Figs. 75, 76. Frons of apt. viv. of *Betulaphis* spp., ventral view. – 75: *pelei* (from Finland); 76: *brevipilosa* (from Denmark); the lateral pairs of frontal hairs are placed on separate tubercles in the former, on transverse ridges in the latter. (Scale 0.25 mm).

70. ***Betulaphis quadrituberculata*** (Kaltenbach, 1843)
Plate 1: 7. Figs. 70, 73.

Aphis quatrituberculata Kaltenbach, 1843: 134. – Survey: 106.

Apterous viviparous female. White, whitish yellow, or pale yellowish green. Tarsi, or apices of tarsi, dark. Abdominal segments with one marginal pair, one pleural pair, and one spinal pair of long, capitate hairs, or the pleural and spinal hairs reduced; marginal hairs on abd. segm. III 1.5–2.7 × IIIbd., spinal hairs on abd. segm. VII 1.3–2.5 × IIIbd. Antenna a little longer than 0.5 × body; processus terminalis as long as VIa or slightly shorter. Apical segm. of rostrum as long as 2sht. or slightly shorter. 1.5–2.0 mm.

Alate viviparous female. Yellowish, sometimes with black dorsal spot. Dorsal hairs pointed, blunt or capitate, those on abd. tergite VI about 2.0 × IIIbd. Antenna about 0.7 × body, a little darker than in apterous viviparous female; segm. III with 8–16 secondary rhinaria (Fig. 73).

Oviparous female. Similar to the apterous viviparous female, but with more or less dark pattern of segmental cross bars on abd. tergites; sometimes shining black over entire dorsum. Posterior part of abdomen prolonged, functioning as an ovipositor. Processus terminalis, tarsi, and hind legs, dark. Apical segm. of rostrum relatively shorter than in apterous viviparous female. Dorsal cuticle a little wrinkled. Hind tibia slightly thickened, with scent plaques.

Apterous male. Yellowish with dark dorsal pattern, reddish in the middle, sometimes black over entire dorsum. Secondary rhinaria on ant. segm. III: 7–18, IV: 1–5, V: 0–5.

Distribution. In Denmark very common and widespread; in Sweden north to Lu.Lpm.; in Norway recorded from several districts north to SFi, mostly far from the coasts; in Finland north to Li. – Widespread in W, C & E Europe, including Iceland, Great Britain (not common), Germany (common), Poland, and NW Russia, south to Hungary; Asia: W Siberia, Mongolia, the Caucasus region; USA and Canada.

Biology. The aphids live on the undersides of leaves of *Betula pubescens* and *B. verrucosa*. In Denmark fundatrices grow up during May and become adults at the end of May or in early June. Both apterous and alate fundatrices are found. The following generations consist mainly of apterae. Sexuales become adults in September and can be found on the branches until November.

Genus *Monaphis* Walker, 1870

Monaphis Walker, 1870: 2001.
Type-species: *Aphis antennata* Kaltenbach, 1843.
Survey: 286.

Only one species in the world.

71. ***Monaphis antennata*** (Kaltenbach, 1843)
Figs. 77–82.

Aphis antennata Kaltenbach, 1843: 115. – Survey: 286.

Alate viviparous female. Green. Antennae black except at base. Body thick, rather strong and compact. Frons with lateral tubercles. Abdomen with marginal tubercles on segm. III–VI. Antenna longer than body, thick, resembling the antenna of a longicorn beetle; processus terminalis nearly twice as long as segm. III, up to about 9 × VIa, at base almost as thick as VIa; segm. III with about 40 small, circular secondary rhinaria on basal $^2/_3$. Rostrum does not reach to 2nd coxae; apical segm. blunt, with up to 6 accessory hairs. First tarsal segm. usually with 9 hairs. Siphunculus very low, almost invisible, with flange. Cauda tongue-shaped (Fig. 80), not constricted, with about 10–12 hairs. 3.3–4.3 mm.

Oviparous female. The only apterous morph. Green. Hind tibia thickened, with about 60–80 scent plaques.

Alate male. Reddish. Smaller and thinner than alate viviparous female, with numerous secondary rhinaria on ant. segm. III–V.

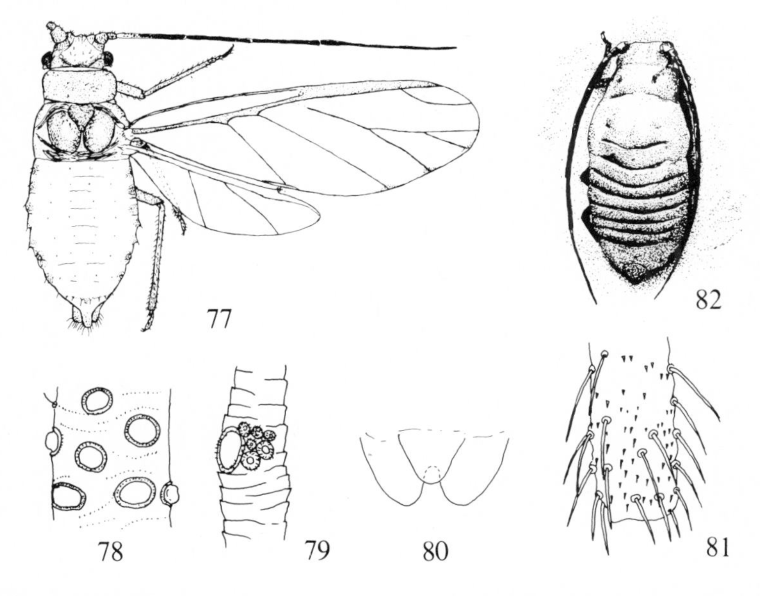

Figs. 77–82. *Monaphis antennata* (Kalt.). – 77: al. viv.; 78: part of ant. segm. III of al. viv. showing the shape of the secondary rhinaria surrounded by rudimentary hairs; 79: part of ant. segm. VI with primary rhinarium and accessory rhinaria; 80: cauda and anal plate of al. viv. in outline, dorsal view; 81: apex of hind tibia of al. viv.; 82: nymph. (77 after Szelegiewicz, redrawn, 82 drawn after photo from Quednau in Börner & Heinze).

Distribution. In Sweden found in several districts, but not commonly, from Sk. north to Dlr.; not in Denmark, Norway, and Finland. – Occurring in the European broad-leaf-region, rather common in the Netherlands, rare in Great Britain and Germany, not known from N Germany; south to Portugal; also known from Hungary, Poland, and NW Russia; W Siberia.

Biology. The host is *Betula verrucosa.* The aphids occur solitarily on the uppersides of leaves, often close to the midrib. Although large insects and freely exposed, they are very difficult to discern. All viviparous females are alate.

SUBTRIBE CALLAPHIDINA

Fore coxae often enlarged, up to about 1.5 times as broad as middle coxae. Large dorsal and marginal tubercles, or finger-like processes, often present.

First instar nymphs with four longitudinal rows of dorsal hairs (i.e. pleural hairs absent); thoracic segments with only one marginal hair on each side; ant segm. II with only one hair; pronotum often partly fused with the head capsule.

The Scandinavian species feed on various trees and bushes (not *Betula* and *Fagus*).

Genus *Callaphis* Walker, 1870

Callaphis Walker, 1870: 2000.
Type-species: *Aphis juglandis* Frisch, 1734
= *Aphis juglandis* Goeze, 1778.
Survey: 120.

All viviparous females alate. Antenna shorter than body; processus terminalis shorter than VIa; segm. III with transverse oval secondary rhinaria; antennal hairs very long. Fore coxae not enlarged. First tarsal segm. with 6–8 hairs. Siphunculus with flange, low, truncate, with 3–4 hairs at base. Cauda knobbed. Anal plate bilobed.

Two species in the world, one species in Europe.

72. ***Callaphis juglandis*** (Goeze, 1778)
Plate 3: 2. Fig. 2.

Aphis juglandis Goeze, 1778: 311. – Survey: 120.

Alate viviparous female. Abdomen yellow with long, broad, brown to blackish cross bars on tergites III–VII, dark dorsal sclerites on other tergites, and dark marginal sclerites. Hind femur with dark band at apex. Siphunculi dark. Cauda rather dark at base. Frons with 5 very small and low tubercles. Antenna about 0.4 × body; segm. III longer than segm. IV+V+VI; processus terminalis 0.5–0.6 × VIa; segm. III with 14–22 rhinaria on basal 80–90%, not on line; longest hair on segm. III 4–6 × IIIbd. Rostrum reaching past 1st coxae; apical segm. slightly shorter than 2sht. Hind tibia of fundatrix sometimes with 1–2 scent plaques. Fore wing with dark anterior edge and dark spots at distal ends of oblique veins; radial sector not as distinct as the other veins; second cubital branch dark-bordered. Distal part of cauda oblong, with about 30–45 hairs. 3.4–4.3 mm.

Oviparous female. Abdomen with dark marginal and dorsal sclerites. Hind tibia more or less thickened, with numerous scent plaques.

Alate male. Much like the alate viviparous female, but more slender. Secondary rhinaria on ant. segm. III: 36–40, IV: 6–8, V: 6.

Distribution. In Denmark known from SJ, EJ, and NEZ; in Sweden known from Sk.; not in Norway or Finland. – Common and widespread in Europe south to Portugal, Spain, Italy, Yugoslavia, and Bulgaria; in Great Britain locally common, but erratic in its appearance; in N Germany common; also known from the Baltic region of Poland and from Russia (but not in NW & W Russia). Asia: Middle East, Caucasus region, C Asia. Western N America.

Biology. The aphids live on the uppersides of leaves of walnut (*Juglans regia*) and are often visited by ants.

Genus *Chromaphis* Walker, 1870

Chromaphis Walker, 1870: 2001.
Type-species: *Lachnus juglandicola* Kaltenbach, 1843.
Survey: 145.

All viviparous females alate. Antenna shorter than body; processus terminalis much shorter than VIa; segm. III with transverse oval, rather broad secondary rhinaria; antennal hairs short. Fore coxae somewhat enlarged. First tarsal segm. normally with 7

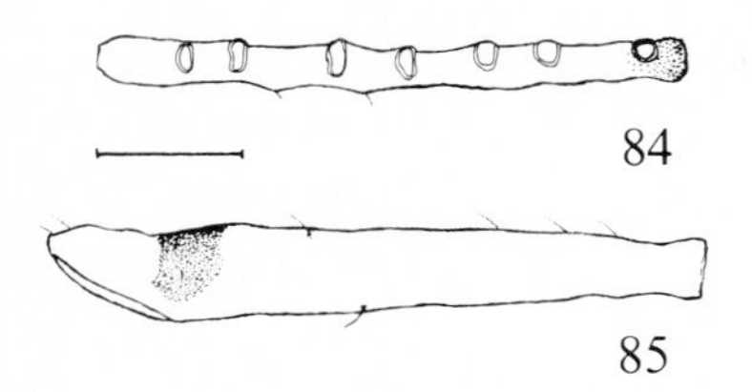

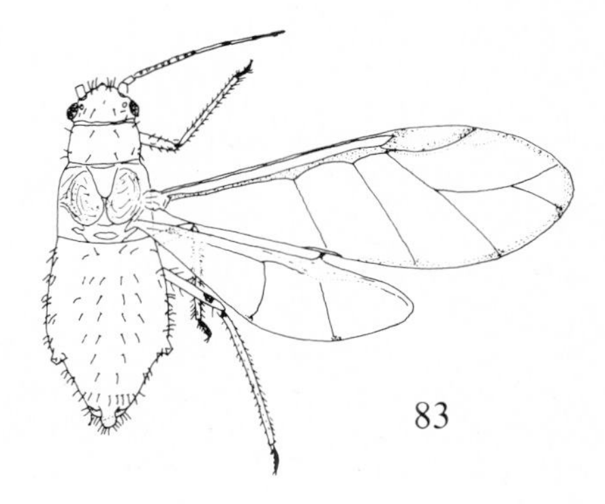

Figs. 83–85. *Chromaphis juglandicola* (Kalt.), al. viv. – 83: body and wings; 84: ant. segm. III; 85: hind femur. (Scale 0.1 mm for 84 and 85). (83 after Szelegiewicz, redrawn, 84 and 85 orig.).

hairs. Siphunculus without flange, low, truncate. Cauda knobbed. Anal plate slightly emarginate. Nymphs with long capitate hairs on frons and margins of abdomen.

Three species in the world, one species in Europe.

73. ***Chromaphis juglandicola*** (Kaltenbach, 1843)
Figs. 83–85.

Lachnus juglandicola Kaltenbach, 1843: 151. - Survey: 145.

Alate viviparous female. Yellowish white, without dark dorsal markings. Apices of ant. segm. and tarsi dark brown. Hind femur with rather large, black spot near apex (Fig. 85); middle femur may have a similar, but smaller spot. Frons with 5 small tubercles. Antenna about 0.5 × body; processus terminalis about 0.25 × VIa or shorter; segm. III with 5–9 secondary rhinaria placed in a row along the entire segment which is a little thickened at level of each rhinarium (Fig. 84); antennal hairs pointed, shorter than 0.5 × IIIbd. Rostrum reaches past fore coxae; apical segm. about 0.75 × 2sht. Wings pale, including pterostigma; fore wing slightly pigmented around bases of media and cubital branches; radial sector weak or absent. Siphunculus conical, smooth, with two ventral hairs. Distal part of cauda transverse, with 15–20 hairs. 1.5–2.6 mm.

Oviparous female. Abd. tergites III–V with dark cross bars. Hind tibia somewhat swollen, with about 35 scent plaques.

Alate male. Very much like the alate viviparous female, but abd. tergites IV–V with paired dark spinal sclerites; marginal sclerites of tergite V dusky. Secondary rhinaria irregularly arranged, on ant. segm. III: 11–24, IV: 5–9, V: 4–7, VI: 2–3.

Distribution. In Denmark known from SJ and NEZ; in Sweden from Sk., Öl., and Gtl.; not in Norway or Finland. - Europe and Asia, east to C Asia, south to Portugal, Spain, and the Middle East; in Great Britain not very common; in N Germany not rare; recorded from Poland, including the Baltic region; not recorded from NW & W Russia. Africa: Atlas Mountains. Introduced in N America.

Biology. The aphids live scattered on the undersides of leaves of *Juglans regia.* In Great Britain the species may occur locally in large numbers and can then cause damage to foliage and immature fruits (Stroyan 1977).

Genus *Myzocallis* Passerini, 1860

Myzocallis Passerini, 1860: 28.
Type-species: *Aphis coryli* Goeze, 1778.
Survey: 290.

All viviparous females of Scandinavian species alate. Frontal tubercles weakly developed. Abdomen without dorsal tubercles, sometimes with marginal tubercles. Each abd. tergite with several spinal hairs. Antenna as long as, or shorter than, body;

processus terminalis longer than VIa; segm. III with rather few transverse oval, rather broad or subcircular, secondary rhinaria forming a row; antennal hairs in adult viviparae short, 0.5–1.0 × IIIbd. Rostrum reaches past fore coxae. Fore coxae only slightly larger than middle and hind coxae. First tarsal segm. normally with 5 ventral and 2 dorsal hairs. Radial sector of fore wing more or less reduced. Siphunculus low, truncate, without flange. Cauda knobbed and anal plate bilobed in viviparous females (Fig. 91). Nymphs with very long capitate dorsal hairs on wart-like elevations.

It is a holarctic genus with 29 species in the world (14 of these in the subgenus *Myzocallis* s. str.), 4 species in Scandinavia. The hosts are deciduous trees and bushes, especially amentaceous trees of various families, e.g. Corylaceae, Fagaceae, and Myricaceae. The aphids are not visited by ants.

Key to species of *Myzocallis*

Alate viviparous females.

1 Abdomen with dark spinal sclerites or cross bars, and dark marginal sclerites. Siphunculi dark. 2
– Abdomen pale, spinal sclerites pale or faintly pigmented. Siphunculi pale. 3
2 (1) Each abdominal sclerite much darker along the edges than in the middle (Fig. 94). Distal end of ant. segm. I as dark as segm. II. .... 77. *myricae* (Kaltenbach)
– Each abdominal sclerite not darker along the edges than in the middle (Fig. 93). Distal end of ant. segm. I distally paler than segm. II. .... 75. *castanicola* Baker
3 (1) Apical segment of rostrum 1.4–1.5 × 2sht. (Figs. 86, 88). .... 76. *coryli* (Goeze)
– Apical segment of rostrum 0.9–1.1 × 2sht. (Figs. 87, 89). .... 74. *carpini* (Koch)

74. ***Myzocallis carpini*** (Koch, 1855)
Figs. 87, 89, 90.

Callipterus carpini Koch, 1855: 216. – Survey: 290.

Alate viviparous female. Colour of body, antennae, and wings usually as in *coryli*, abdomen pale, spinal and marginal sclerites not or at most faintly, pigmented. Ant. segm. III with 2–6 rhinaria; processus terminalis about 2.3–3.0 × VIa. Apical segm. of rostrum 0.9–1.1 × 2sht. (Figs. 87, 89). Radial sector reduced. 1.3–2.2 mm.

Distribution. In Denmark common and widespread; in Sweden north to Upl.; in Norway known from AK (trap); not in Finland. – Europe; in Great Britain widespread but not very common, in N Germany not rare; known from Poland, but not NW & W Russia; France, Hungary, and Caucasus. Introduced in N America (British Columbia).

Biology. The aphids are found on the undersides of leaves of *Carpinus betulus*.

75. ***Myzocallis castanicola*** Baker, 1917
Fig. 93.

Myzocallis castanicola Baker, 1917: 424. – Survey: 290.

Alate viviparous female. Yellow with black longitudinal stripes on head and pronotum and black spots on abdomen. Antennae dark, except for segm. I and bases of segm. III and IV. Apices of legs and siphunculi dark. Secondary rhinaria on ant. segm. III: 5–7. Processus terminalis 1.5–2.0 × VIa. Apical segm. of rostrum about 1.0 × 2sht. Radial sector of fore wing well developed. 1.6–1.7 mm.

Distribution. In Denmark known from EJ, NWJ, and NEZ; in Sweden known from Sk., Bl., and Upl.; not in Norway or Finland. – Europe, south to Portugal and Spain, east to Russia and Caucasus; in Great Britain widely distributed, but local; in N Germany not rare; known from Poland, including the Baltic region. Africa: St. Helena, South Africa. Australia, New Zealand. N America: California, British Columbia. S America: Argentina, Brazil.

Biology. The aphids live on the undersides of leaves of *Castanea sativa,* occasionally also oak (*Quercus* spp.).

76. ***Myzocallis coryli*** (Goeze, 1778)
Plate 3: 4. Figs. 86, 88, 91, 92.

Aphis coryli Goeze, 1778: 311. – Survey: 290.

Alate viviparous female. Yellowish white. Ant. segm. IV, V, and VIa with dark apices; segm. III with 3–5 rhinaria on basal half; processus terminalis about 2.2–2.7 × VIa. Apical segm. of rostrum 1.4–1.5 × 2sht. (Figs. 86, 88). Fore wing with black spot in basal part of pterostigma; radial sector weakly developed. 1.3–2.0 mm.

Oviparous female. Pale. With very long capitate hairs. Antennae without secondary rhinaria; segm. III with long capitate hair, 1.3–1.5 × IIIbd. Basal half of hind tibia strongly swollen, pale, with numerous scent plaques.

Alate male. Head and thorax brown, abdomen with dark dorsal cross bars. Secondary rhinaria on ant. segm. III: 13–20, IV: 3–7, V: 2–6, VIa: 1–3.

Distribution. In Denmark very common and widespread; in Sweden common in the southern part of the country, north to Dlr. and Upl.; in Norway north to SFi and On; in Finland known from the southern part of the country (Ab, N, Ta, Sa). – Widespread in Europe, from the British Isles east to Russia, south to the Mediterranean Sea; very common in Great Britain, N Germany, and Poland; also in NW & W Russia. Asia: Turkey, Caucasus. Africa: Atlas Mountains. Introduced in Australia (Tasmania), New Zealand, and America; widespread in the USA and Canada; recorded also from Argentina and Chile.

Biology. The aphids live on the undersides of the leaves of *Corylus avellana.* Sexuales occur in Denmark in October.

77. ***Myzocallis myricae*** (Kaltenbach, 1843)
Plate 3: 5. Fig. 94.

Aphis myricae Kaltenbach, 1843: 96. – Survey: 291.

Alate viviparous female. Yellow or orange, with black longitudinal stripes on head and pronotum, and spots on abdomen; each spot rather paler centrally. Antennae rather dark with darker apices of segments. Siphunculi brown. Ant. segm. III with 3–5 rhinaria; processus terminalis 1.2–1.8 × VIa. Apical segm. of rostrum about 1.0 × 2sht. Radial sector not reduced; wings sometimes shortened. 1.8–2.0 mm.

Distribution. In Denmark very common in Jutland and on Læsø; in Sweden common, from Bl. in the south to Vb. in the north; in Norway in districts along the southern

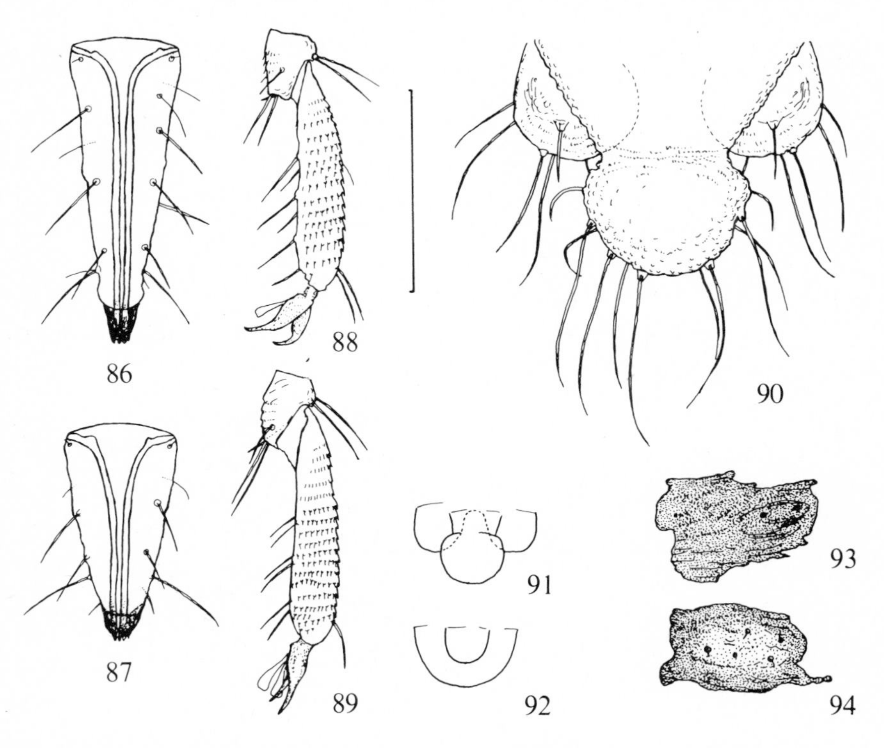

Figs. 86–94. *Myzocallis* spp. – 86, 87: apical segment of rostrum of al. viv. of *coryli* (Goeze) (86) and *carpini* (Koch) (87); 88, 89: hind tarsus of al. viv. of *coryli* (88) and *carpini* (89); 90: cauda and anal plate of *carpini*, dorsal view; 91, 92: cauda and anal plate of *coryli* in outline, dorsal view; 91: al. viv.; 92: ovip. female; 93, 94: dorsal sclerite on left side of abd. tergite IV of al. viv. of *castanicola* Baker (93) and *myricae* (Kalt.) (94). (Scale 0.1 mm for 86–90).

coast, north to MRy; in Finland in Ta and Sa. – Europe; widely distributed in the north and west of the British Isles; common in N Germany; also known from the Baltic coast of Poland and in NW Russia.

Biology. The aphids live on leaves of *Myrica gale,* especially on the young shoots.

Genus *Tuberculatus* Mordvilko, 1894 s. lat.

Tuberculatus Mordvilko, 1894: 51.
Type-species: *Aphis quercea* Kaltenbach, 1843.
Survey: 439.

All viviparous females alate. Frons with low tubercles. Antenna as long as, or longer than, body; processus terminalis a little shorter, as long as, or longer than, VIa; segm. III with rather broad, transverse oval or subcircular, secondary rhinaria. Fore coxae a

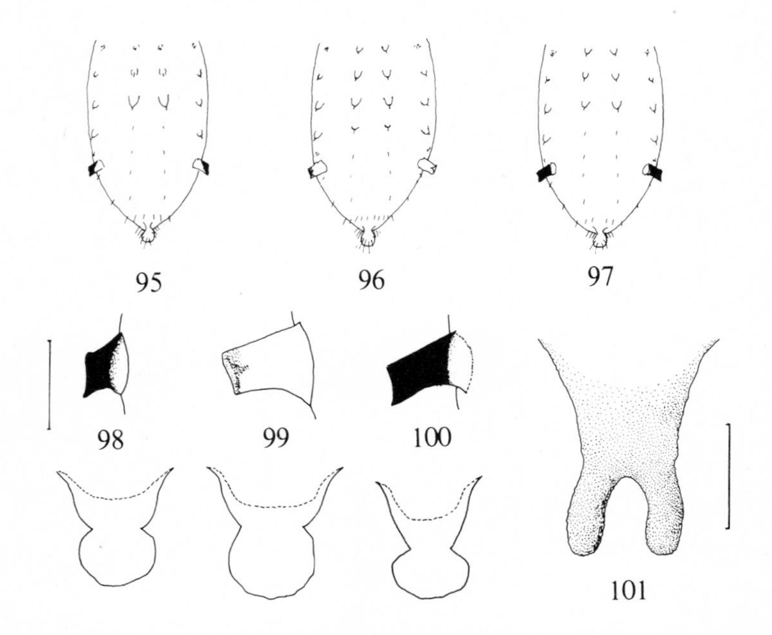

Figs. 95–97. Abdomen of al. viv. of *Tuberculatus (Tuberculoides)* spp. – 95: *annulatus* (Hartig); 96: *borealis* (Krzywiec); 97: *neglectus* (Krzywiec). (After Stroyan, redrawn).
Figs. 98–100. Siphunculus (above) and cauda (below) (hairs omitted). – 98: *annulatus;* 99: *borealis;* 100: *neglectus.* (Scale 0.1 mm). (After Stroyan, redrawn).
Fig. 101. Dorsal tubercle of *Tuberculatus querceus* (Kalt.). (Scale 0.1 mm).

little larger than middle and hind coxae. First tarsal segm. with 5–6 ventral and 2 dorsal hairs. Abdomen with dorsal and marginal tubercles. Siphunculi short, truncate, smooth. Cauda knobbed. Anal plate bilobed. Nymphs with long capitate hairs.

There are 37 species in the world. The genus is subdivided into six subgenera. The Scandinavian species belong to two of them.

Key to subgenera of *Tuberculatus*

Alate viviparous females

1 Abdomen with two large, usually basally fused, spinal processes on tergite III (Fig. 101). Body wax-covered. Antenna longer than 1.3 × body. .... *Tuberculatus* Mordvilko s.str. (p. 53)
– Abdomen with paired spinal processes (or tubercles) on more than one tergite (Figs. 95–97). Body not wax-covered. Antenna shorter than 1.3 × body. .... *Tuberculoides* van der Goot (p. 54)

Subgenus *Tuberculatus* Mordvilko, 1894 s. str.

Frons with weakly developed tubercles. Antenna much longer than body. Abd. tergite III with a large V- or Y-shaped dorsal process formed by the spinal tubercles. Spinal tubercles absent from other tergites.

Three species in the world, one species in Scandinavia.

78. ***Tuberculatus (Tuberculatus) querceus*** (Kaltenbach, 1843)
Fig. 101.

Aphis quercea Kaltenbach, 1843: 136. – Survey: 439.

Alate viviparous female. Dirty greenish, with white wax-powder. Head, apices of ant. segm., thorax, spot on distal part of femur, tarsi, abd. tubercles, and siphunculi brownish to black. Spinal tubercles on abd. segm. III fused at base, 2–3 times as long as siphunculus, finger-shaped (Fig. 101); abdomen also with 1–3 finger-shaped marginal tubercles. Antenna more than 1.3 × body; processus terminalis 1.3–1.7 × VIa; segm. III with 6–14 secondary rhinaria in a row along entire segment. Rostrum reaches between fore and middle coxae; apical segm. of rostrum about 0.75 × 2sht. Fore wing with two brown spots in pterostigma; radial sector sometimes weakly developed. Fore coxae enlarged. 1.4–2.4 mm.

Oviparous female. The only apterous morph. Thorax and abd. segm. I–IV with dark spinal and marginal sclerites; spinal tubercles on abd. segm. III very low, not fused. Ant. segm. III with 2–7 secondary rhinaria. Subsiphuncular wax gland plates absent. Posterior part of abdomen prolonged. Hind tibiae swollen, with numerous scent plaques.

Alate male. Spinal process on abd. segm. III smaller than in the alate viviparous female. Secondary rhinaria present on ant. segm. III, IV, V, and VIa.

Distribution. In Denmark known from EJ, NWJ, and NEZ, rather rare; in Sweden found in Sk., Hall. and Öl.; in Norway known from AK and HO; not in Finland. – Europe; in Great Britain local and rather uncommon; in N Germany apparently rare; recorded from Poland (but not from the Baltic region) and W Russia. South to Turkey and Caucasus.

Biology. The aphids are found scattered on the undersides of leaves of *Quercus robur*. They are not visited by ants.

Subgenus *Tuberculoides* van der Goot, 1915

Tuberculoides van der Goot, 1915: 312.
Type-species: *Aphis quercus* Kaltenbach, 1843
= *Aphis annulata* Hartig, 1841.
Survey: 441.

Frons with 5 low tubercles. Antenna about as long as body, or a little longer. Abd. tergites I–III(–IV) with paired spinal processes (or tubercles) (Figs. 95–97).

Nine species in the world, three species in Scandinavia. They feed on oak and are not visited by ants. The key below is based on Stroyan (1977).

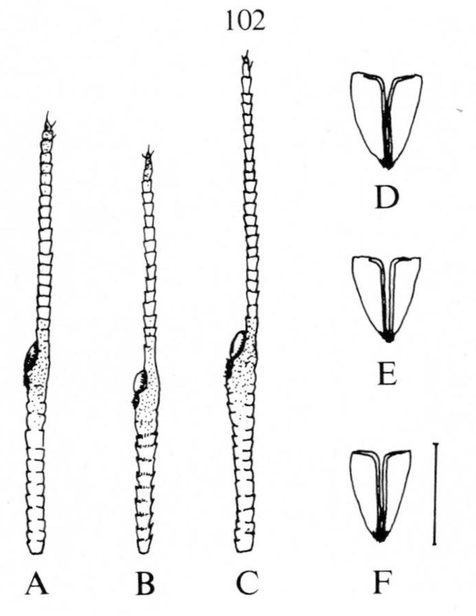

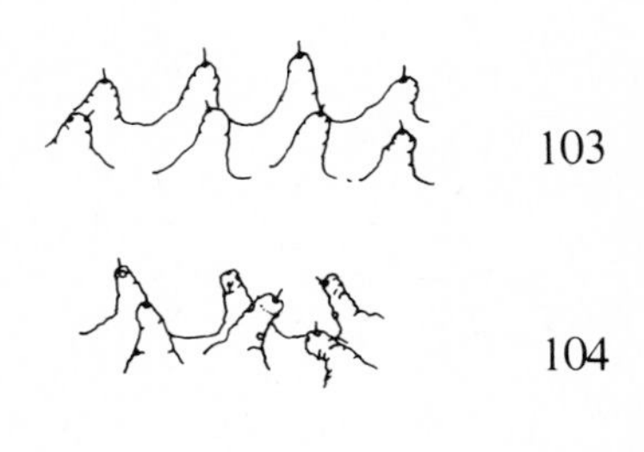

Fig. 102. Ant. segm. VI (A B, C) and apical segm. of rostrum in outline (D, E, F) of *Tuberculatus (Tuberculoides)* spp. A & D: *annulatus;* B & E: *borealis;* C & F: *neglectus.* (Scale 0.1 mm). The length of hind tibia of the same specimens: *annulatus* 1.08 mm, *borealis* 1.14 mm, *neglectus* 1.00 mm.

Figs. 103–104. Abdominal spinal processes, lateral view. – 103: *borealis;* 104: *neglectus.* (After Krzywiec, redrawn).

Key to species of subgenus *Tuberculoides*

Alate viviparous females

1 Spinal tubercles on abd. segm. I–II much smaller than those on segm. III, sometimes inconspicuous; spinal tubercles on segm. III sometimes dusky; segm. IV without spinal tubercles (Fig. 95). .... 79. *annulatus* (Hartig)

– Spinal tubercles on abd. segm. I–III of rather even size, pale; segm. IV sometimes with smaller spinal tubercles. 2

2 (1) With values above 2.0 for the function a^2/bc, when a = length of ant. segm. VI, b = length of hind tibia, and c = length of apical segm. of rostrum (Fig. 102 C & F). Spinal tubercles on abd. segm. IV usually absent (Fig. 97). .... 81. *neglectus* (Krzywiec)

– With values below 2.0 for the function a^2/bc (see above) (Fig. 102 B & E). Very small spinal tubercles on abd. segm. IV usually present (Fig. 96). .... 80. *borealis* (Krzywiec)

79. ***Tuberculatus (Tuberculoides) annulatus*** (Hartig, 1841)
Plate 3: 6. Figs. 95, 98, 102 A & D.

Aphis annulatus Hartig, 1841: 369. – Survey: 441.

Alate viviparous female. Yellowish, greyish green, or pink. Apices of ant. segm., tarsi, and distal two thirds or more (of anterior margin) of siphunculi dark. Abdomen with one pair of rather large, often dusky, spinal tubercles or processes on tergite III; tergites I and II each with one pair of less developed spinal tubercles (Fig. 95). Wart-like marginal tubercles present on some of the abd. segm. Processus terminalis 0.7–1.1 × VIa (Fig. 102A); ant. segm. III with 3–12 rhinaria in a row on basal ½–⅔. Rostrum reaches past fore coxae; apical segm. of rostrum 0.08–0.10 mm long, shorter than 2sht., with 5–8 accessory hairs. Radial sector of fore wing sometimes weakly developed. Greatest length of siphunculus 0.04–0.07 mm, or at most equal to length of caudal knob, but usually shorter than this (Fig. 98). 1.7–2.2 mm.

Oviparous female. Pale. Siphunculi pale. Abd. tubercles reduced. Antennae without secondary rhinaria. Basal two thirds of hind tibia thickened, pale, with many indistinct scent plaques. Cauda and anal plate rounded, semicircular.

Alate male. Head, thorax, antennae, and legs darker than in the alate female. Abd. segments with dark dorsal cross bars and marginal sclerites. Siphunculi black. Spinal tubercles reduced; marginal tubercles on abd. segm. IV well developed. Secondary rhinaria on ant. III: about 55, scattered all over the segment, IV: about 20–23, V: about 8–11, VIa: 2–4.

Distribution. In Denmark extremely common all over the country; in Sweden common from Sk. in the south to Vb. in the north; in Norway common along the coast of the southern part of the country north to HOy; in Finland north to Obs. – Europe and Asia, east to W Siberia and C Asia, south to Portugal, Spain, Turkey, and the Caucasus region; very common in the British Isles, N Germany, Poland, and NW & W Russia. Also known from N Africa. Introduced in Australia, New Zealand, N America and S America.

Biology. The aphids live on the undersides of leaves of oak, often abundantly on *Quercus robur*, less frequently on *Q. petraea* or other *Q.* spp. They are scattered on the leaves or in small colonies, and are not visited by ants. In Denmark the sexuales are found in October.

80. ***Tuberculatus (Tuberculoides) borealis*** (Krzywiec, 1971)
Figs. 96, 99, 102 B & E, 103.

Tuberculoides borealis Krzywiec, 1971: 327. – Survey: 441.

Alate viviparous female. Rather similar to *neglectus*. Body colour mottled green and yellowish. Siphunculus pale or blackish on distal 50% (of anterior margin) or less (seldom up to 65%). Abdomen with paired spinal tubercles on segm. I–IV, those on I–III of rather even size, those on IV sometimes very small (Figs. 96, 103) or apparently absent. Antenna 0.8–1.0 × body; processus terminalis 0.9–1.3 × VIa (Fig. 102B); segm. III with 3–7 rhinaria. Apical segm. of rostrum 0.09–0.12 mm, 0.8–0.9 × 2sht., with 6–11 accessory hairs. Greatest length of siphunculus 0.06–0.10 mm, 0.8–1.2 × the length of the caudal knob (Fig. 99). 1.9–2.3 mm.

Distribution. In Denmark known from NEZ (suction trap at Tåstrup near Copenhagen, according to unpublished information from Rothamsted Exp. Stat., Harpenden, England); in Sweden found in Sk. (Danielsson in litt.); not in Norway or Finland. – Known from Great Britain and N Poland. Krzywiec suggests that the southern boundary runs through the Pomeranian district of Poland. It is probably widespread in N Europe, but overlooked.

Biology. The aphids occur scattered on leaves of *Quercus robur*, more rarely on *Q. petraea* and other *Q.* spp. The species has probably been overlooked in many districts, because it lives on the same host as *annulatus* and can easily be mistaken for this very common species.

81. ***Tuberculatus (Tuberculoides) neglectus*** (Krzywiec, 1966)
Figs. 97, 100, 102 C & F, 104.

Tuberculoides neglectus Krzywiec, 1966: 595. – Survey: 441.

Alate viviparous female. Yellowish. Mesothorax yellow. Apices of ant. segm., tarsi, and distal 55–95% (of anterior margin) of siphunculi, brown or black. Abdomen with paired spinal tubercles of rather even size on segm. I–III (Fig. 97, 104). Wart-like marginal tubercles present on abd. segm. I–VII. Antenna usually a little longer than body; processus terminalis 1.0–1.6 × VIa (Fig. 102 C); ant. segm. III with 3–7 rhinaria in a row on basal part. Rostrum reaches past fore coxae; apical segm. of rostrum 0.08–0.10 mm, 0.7–0.9 × 2sht., with 4–9 accessory hairs. Radial sector weakly developed. Greatest length of siphunculus 0.07–0.11 mm, 1.3–2.1 × length of caudal knob (Fig. 100). 1.7–2.3 mm.

Distribution. In Denmark found in EJ (near Vrads); the species has according to Stenseth (1979 in litt.) also been found in Norway; not yet found in Sweden and Finland, but probably overlooked. – Known from Great Britain, the Netherlands, Poland, Austria, Czechoslovakia, and Hungary.

Biology. The aphids live on the undersides of leaves of *Quercus petraea,* rarely on *Q. robur* or hybrid *Q. robur* × *petraea.* Sexuales have been observed in November in Poland (Krzywiec 1966).

Genus *Pterocallis* Passerini, 1860

Pterocallis Passerini, 1860: 28.
Type-species: *Aphis alni* Fabricius, 1781
= *Aphis alni* DeGeer, 1773.
Survey: 367.

Viviparous females may be apterous or alate. Nymphs and apterous adults with numerous long, blunt or capitate, dorsal hairs, which look serrate or rough because of minute denticles (Figs. 110, 111). Abdomen without spinal processes, but sometimes in alate females with indistinct median convexities on the anterior tergites. Marginal tubercles present on pronotum and abd. segments, and are divided into small wax-producing facets. Antenna as long as body or shorter; processus terminalis shorter than VIa; secondary rhinaria circular or subcircular, present on ant. segm. III of alatae, usually absent from apterae. Fore coxae slightly enlarged, more distinctly so in alatae than in apterae. First tarsal segm. with 5 ventral and 0–2 dorsal hairs. Radial sector of fore wing more or less obsolete. Siphunculi low, truncate. Cauda knobbed. Anal plate bilobed in viviparae.

There are 17 species of *Pterocallis* s. lat. in the world, three species in Scandinavia. The genus is subdivided into four subgenera. The Scandinavian species all belong to *Pterocallis* s. str. Their host plant is *Alnus.* One species is visited by ants, two are not.

Key to species of *Pterocallis*

Apterous viviparous females

1 Apices of ant. segments not black. Processus terminalis 0.2–

0.3 × VIa. Body colour white. .... 82. *albidus* Börner

– Apices of ant. segments black. Processus terminalis 0.5–0.8 × VIa. Body colour yellowish white, yellowish, or greenish. 2

2 (1) Yellowish white og yellowish green. Ant. segm. III with 1–2 long hairs on inner side of basal half (Fig. 108); all other antennal hairs much shorter and inconspicuous. Dorsal body hairs pale. .... 83. *alni* (DeGeer)

– Yellowish or green with darker green dorsal markings. Ant. segm. III usually with 3 long hairs on inner side, the most distal placed on the distal half of the segm. (Fig. 109); all other antennal hairs somewhat shorter, but conspicuous. Dorsal body hairs pigmented. .... 84. *maculatus* (v. Heyden)

Alate viviparous females

1 Ant. segm. III with 6–9 secondary rhinaria. .... 82. *albidus* Börner

– Ant. segm. III with 2–5 secondary rhinaria. 2

2 (1) Flagellum (= ant. segm. III–VI) 10–11 × the length of apical segm. of rostrum. Ant. segm. VI 7.5–9 times as long as its maximal width. .... 84. *maculatus* (v. Heyden)

– Flagellum 15–17 × apical segm. of rostrum. Ant. segm. VI 9.5–12.5 times as long as its maximal width. .... 83. *alni* (DeGeer)

82. ***Pterocallis albidus*** Börner, 1940
Plate 4: 2. Fig. 105.

Pterocallis albida Börner, 1940: 2. – Survey: 367.

Apterous viviparous female. Very much like *alni*, but paler, white. Apices of ant. segm. not black. Dorsal hairs capitate. Antenna 0.7–0.8 × body; processus terminalis 0.2–0.3 × VIa. Apical segm. of rostrum about 0.75 × 2sht. About 1.2 mm.

Alate viviparous female. Very much like the alate viviparous female of *alni*, but ant. segm. III with more secondary rhinaria, 6–9.

Distribution. In Denmark known from WJ and NWJ; in Sweden known from Sk. and Vb.; not in Norway; in Finland known from Ta and Ok. – Probably widespread and common in Europe, including Scandinavia, but overlooked; described from Germany; also known from Poland, Czechoslovakia, and Baltic region of the USSR. There are no records from Great Britain or N Germany.

Biology. The aphids live scattered on the undersides of leaves of *Alnus incana*. They are not visited by ants.

83. ***Pterocallis alni*** (DeGeer, 1773)
Plate 4: 1. Figs. 106, 108, 110.

Aphis alni DeGeer, 1773: 47. – Survey: 367.

Apterous viviparous female. Yellowish white or yellowish green. Apices of ant. segm. and tarsi black. Hind femur with black spot on outer side near the knee. Siphunculi dark with pale bases. Dorsal body hairs pale (Fig. 110). Antenna 0.6–0.8 × body; flagellum (= segm. III–VI) about 1.5 × hind tibia; processus terminalis about 0.5–0.8 × VIa; segm. III with 1–2 longer hairs on inner side of basal half (Fig. 108); other antennal hairs very short. Apical segm. of rostrum about 0.8–0.9 × 2sht. 1.5–2.0 mm.

Alate viviparous female. Apices of ant. segm. black. Hind femur with black spot as in aptera. Siphunculi dark with pale bases. Hairs fine, pointed. Antennal hairs in-

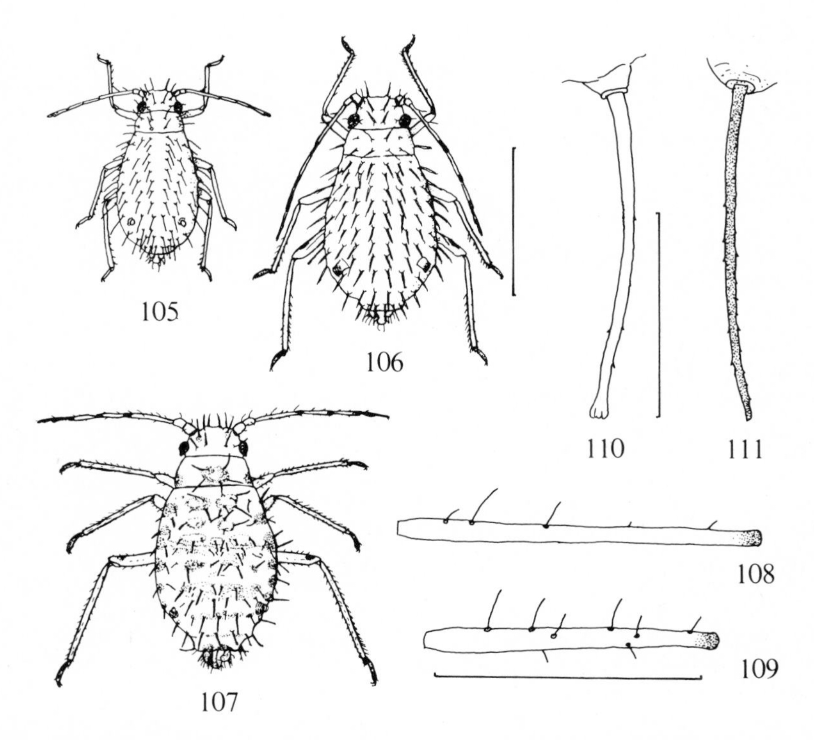

Figs. 105–111. *Pterocallis* spp. – 105: *albidus* Börner, apt. viv.; 106: *alni* (DeGeer), apt. viv.; 107: *maculatus* (v. Heyd.), apt. viv.; 108, 109: ant. segm. III of apt. viv. of *alni* (108) and *maculatus* (109); 110, 111: dorsal body hair of *alni* (110) and *maculatus* (111). (Scales 1 mm for 105 and 106, 0.5 mm for 108 and 109, 0.1 mm for 110 and 111). (107 after Szelegiewicz, redrawn, others orig.).

conspicuous, up to about 1.0 × IIIbd. Ant. segm. III with 2–5 rather large rhinaria on basal half. Radial sector of fore wing absent, or the distal part present; second branch of cubitus darker than the other veins, at base surrounded by a triangular dark spot.

Oviparous female. Yellow. Very much like the apterous viviparous female. Subsiphuncular wax gland plates present, consisting of triangular facets. Hind tibiae strongly swollen, with numerous scent plaques. Anal plate rounded, not bilobed.

Alate male. Rather similar to the alate female. Ant. segm. III with 11–14 secondary rhinaria in a row along the whole segment, IV with 5–7, V and VIa also with a few secondary rhinaria.

Distribution. In Denmark very common all over the country; in Sweden common and widespread, north to Nb.; in Norway common and widespread, north to TRi; in Finland recorded from Ab, N, St, and ObS. – Widespread in Europe, from the British Isles in the west to the USSR in the east, south to Portugal and Spain; in Great Britain widespread, but rarely very abundant; in N Germany very common; also occurring in the Baltic region of Poland. Asia: Turkey. Introduced in N America, widespread in the USA and Canada. According to Ilharco (1973) also in New Zealand.

Biology. The aphids live on the undersides of leaves of *Alnus glutinosa,* not in colonies, but more or less scattered. They are not visited by ants. Sexuales can be found in October.

84. ***Pterocallis maculatus*** (von Heyden, 1837)
Figs. 107, 109, 111.

Aphis maculata von Heyden, 1837: 297. – Survey: 367.

Apterous viviparous female. Green or yellowish with more or less distinct pattern of dark green transverse bands on dorsum. Apices of ant. segm. and tarsi black. Hind femur with black spot near the knee. Distal end of siphunculus dark. Dorsal body hairs pigmented (Fig. 111), placed on pale, inconspicuous sclerites. Antenna 0.4–0.6 × body; processus terminalis about 0.5 × VIa; segm. III with 0–2 secondary rhinaria, usually with 3 long hairs on inner side, the most distal hair placed on distal half of the segment (Fig. 109); other antennal hairs somewhat shorter, however, rather strong and conspicuous. Apical segm. of rostrum about as long as 2sht. 1.4–2.1 mm.

Alate viviparous female. Abdomen greenish with green markings. Very much like the alate viviparous female of *alni.* The differences are given in the key.

Oviparous female. Light green, with dark transverse dorsal bands on abdomen. Dorsal body hairs long and black. Basal half or two thirds of hind tibia slightly swollen, with about 50 scent plaques.

Apterous male. Brownish, with dark transverse bands on abdomen. Dorsal body hairs long, black. Antenna about 0.75 × body; secondary rhinaria on segm. III: 6–9, IV: 2–4, V: 2–4.

Distribution. In Denmark known from SJ (Flensborg Fjord) and NWJ (Højslev at

Skive), apparently very rare; in Sweden known from Öl. and Upl.; not in Norway; in Finland known from Ta and Sa. - Europe, rather widespread, but not common, south to Portugal and Hungary, east to the western part of the USSR; in Great Britain and N Germany rare; known from Poland (including the Baltic region) and NW & W Russia.

Biology. The aphids occur in colonies on *Alnus glutinosa,* especially along veins on the undersides of the leaves. They are visited by ants.

Genus *Ctenocallis* Klodnitzki, 1924

Ctenocallis Klodnitzki, 1924: 61 (as *Ctenocallis* Grossheim in litt.).
Type-species: *Ctenocallis dobrovljanskyi* Klodnitzki, 1924.
Survey: 166.

Viviparous females apterous or alate. All body segments of apterous females with finger-shaped processes. Antenna shorter than body; processus terminalis shorter than VIa; secondary rhinaria transverse oval, present in all morphs. Siphunculi pore-shaped, placed on marginal processes on abd. segm. VI. Cauda knobbed. Anal plate bilobed in viviparae.

Three species in the world, one species in Scandinavia. The hosts are leguminous plants.

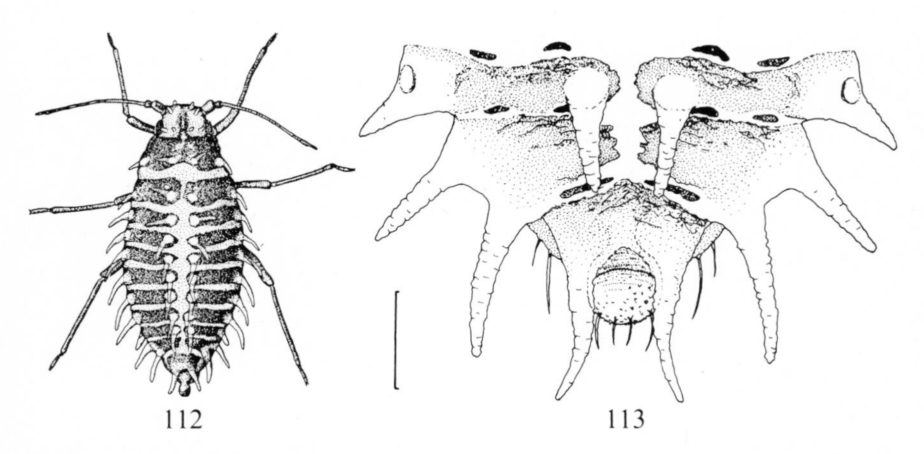

Figs. 112, 113. *Ctenocallis setosus* (Kalt.), apt. viv. - 112: habitus; 113: posterior part of abdomen. The siphuncular pores are placed at bases of marginal processes of abd. segm. VI. (Scale 0.1 mm for 113). (112 after Stroyan, Hille Ris Lambers del.; 113 orig.).

85. ***Ctenocallis setosus*** (Kaltenbach, 1846)
Plate 4: 3. Figs. 112, 113.

Aphis setosa Kaltenbach, 1846: 172. – Survey: 166.

Apterous viviparous female. Yellowish, with brown dorsal cross bars on thorax and abdomen; those on pronotum and abd. segm. VIII entire, the remainder interrupted in the middle. All body segments with long, finger-shaped, mostly backwardly directed processes, one spinal pair and one marginal pair on each of the segments from prothorax to abd. segm. VI, one pleural pair and one marginal pair on abd. segm. VII, one spinal pair on abd. segm. VIII; head with four pairs of shorter processes. Antenna 0.4 × body; processus terminalis about 0.5 × VIa; mid section of segm. III with about 3 secondary rhinaria. Apical segm. of rostrum a little shorter than 2sht. Siphuncular pores placed anteriorly on basal half of marginal process of abd. segm. VI (Fig. 113). 1.4–1.8 mm.

Alate viviparous female. Abdomen yellowish, with four longitudinal rows of dorsal sclerites instead of cross bars (as in the apterous viviparous female). Dorsal processes smaller than in the apterous viviparous female, conical. Radial sector more or less obsolete; second branch of cubitus ending in a dusky spot at posterior margin of fore wing.

Oviparous female. Similar to the apterous viviparous female, but dorsal processes on posterior abd. segments smaller. Subsiphuncular wax gland plates absent. Ant. segm. III with 1–2 rhinaria. Hind tibiae swollen, with numerous, rather indistinct, scent plaques on basal two thirds. Anal plate not bilobed.

Alate male. Similar to the alate viviparous female, but smaller, about 1 mm. Antenna longer than 0.75 × body; secondary rhinaria on segm. III: about 9, IV: 1, V: 0–1, VIa: 1.

Distribution. In Denmark one record from EJ (Femmøller); not in Sweden, Norway, or Finland. – Apparently rare in NW & C Europe, east to Poland, south to Languedoc in France; rare in Great Britain and N Germany. In N America recorded from North Carolina and from the Pacific region.

Biology. The aphids live on the uppersides of leaves of *Sarothamnus scoparius*, flattened against the midribs. They are not visited by ants.

Genus *Tinocallis* Matsumura, 1919

Tinocallis Matsumura, 1919: 100.
Type-species: *Tinocallis ulmiparvifoliae* Matsumura, 1919.
Survey: 428.

All viviparous females alate. Abdomen with dorsal and marginal tubercles. Antenna as long as body, or shorter; processus terminalis shorter than VIa; secondary rhinaria transverse oval, narrow (Fig. 114), present only on segm. III. Fore coxae considerably larger than middle and hind coxae (Fig. 116). First tarsal segm. with 7 or 8 hairs. Radial sector of fore wing present or absent. Siphunculus stump-shaped, broad at base, smooth. Cauda knobbed. Anal plate bilobed. Nymphs with long dorsal and marginal hairs which are capitate or cleft at apex.

There are 16 species in the world, two species in Scandinavia. They feed on *Ulmus* and are not attended by ants. They are able to jump by means of muscles in the greatly enlarged fore coxae.

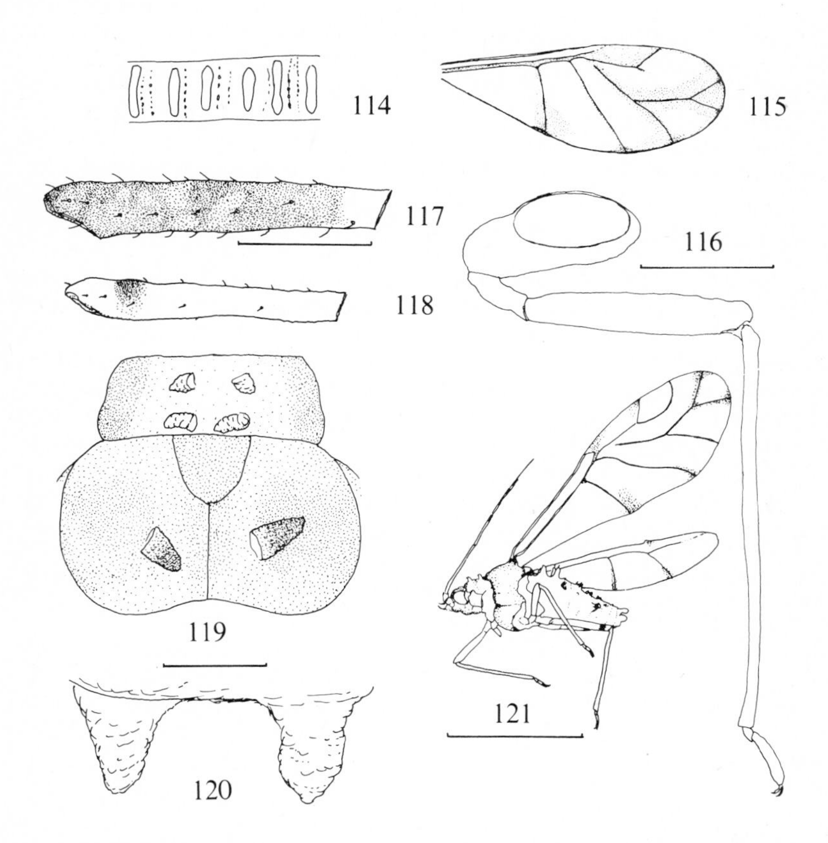

Figs. 114–117. *Tinocallis platani* (Kalt.), al. viv. – 114: part of ant. segm. III with secondary rhinaria; 115: fore wing; 116: fore leg with enlarged coxa (hairs and pigmentation not drawn); 117: hind femur.

Figs. 118–121. *T. saltans* (Nevsky), al. viv. – 118: hind femur; 119: pronotum and part of messonotum with finger-shaped processes; 120: finger-shaped spinal processes on abd. tergite II; 121: body and wings, lateral view. (Scales 0.25 mm for 116, 0.5 mm for 117 and 118, 0.1 mm for 120, 1 mm for 121). (115 after Szelegiewicz, 121 after Nevsky, redrawn; others orig.).

Key to species of *Tinocallis*

Alate viviparous females

1 Pro- and mesonotum with finger-shaped processes (Fig. 119). Processus terminalis about 0.9 × VIa. Apical segm. of rostrum as long as 2sht., or shorter. .... 87. *saltans* (Nevsky)
– Pro- and mesonotum without finger-shaped processes. Processus terminalis about 0.2 × VIa. Apical segm. of rostrum 1.2–1.5 × 2sht. .... 86. *platani* (Kaltenbach)

86. ***Tinocallis platani*** (Kaltenbach, 1843)
Figs. 114–117.

Lachnus platani Kaltenbach, 1843: 152. – Survey: 428.

Alate viviparous female. Yellow or greenish white, with black or brown markings. Head and pronotum with three dark longitudinal stripes; apices of antennal segments, hind femur (except at base) (Fig. 117), basal part of hind tibia, all tarsi, and siphunculi, more or less darkened. Abd. tergites III–VI with irregularly shaped dark spots, or cross bars, and marginal sclerites. Frontal tubercles weakly developed. Spinal tubercles on abd. segm. I–II finger-shaped, rather short, pale at base, with darker rounded distal part, each provided with a hair. Antenna about 0.75 × body; processus terminalis about 0.2 × VIa; ant. segm. III with 5–20 rhinaria in a row on basal ½–⅔; antennal hairs shorter than 0.5 × IIIbd.; all segments sculptured with small spinules (Fig. 114). Rostrum reaches past fore coxae; apical segm. 1.2–1.5 × 2sht. Fore wings with radial sector reduced, dark shadows present, and surrounding media, second branch of cubitus, and apex of first branch of cubitus (Fig. 115). – Hairs of nymphs terminating in short furca, black, placed on pale tubercles. – 2.0–2.2 mm.

Distribution. In Denmark rare, known from NWJ and NEZ; in Sweden known from Sk. and Gtl.; in Norway from AK and Os; in Finland from N and Ta. – Europe and Asia, east to Siberia, C Asia, and Mongolia, south to Spain, Hungary, Caucasus, and Kazakhstan; rare in the British Isles and N Germany; known from Poland, and NW & W Russia. Introduced in western N America.

Biology. The aphids live on the undersides of leaves of *Ulmus* spp., especially *U. laevis*. Large numbers may be found on a single tree, even in areas where the species is rare.

87. ***Tinocallis saltans*** (Nevsky, 1929)
Figs. 118–121.

Tuberocallis saltans Nevsky, 1929b: 221. – Survey: 429.

Alate viviparous female. Yellow. Rather similar to *platani,* but paler. Pronotum with dark margins. Apices of ant. segm. dark. Hind femur pale, with dark spot near apex (Fig. 118). Siphunculi dusky. Pronotum with 2 spinal pairs of pale, finger-shaped processes; mesonotum with one pair of dark, finger-shaped processes (Fig. 119). Abdomen with pale, finger-shaped, but rather short, spinal processes on segm. I and II (Fig. 120), other segments with dark wart-like hairy dorsal tubercles, and pale or dusky marginal tubercles. Antenna 0.8–0.9 × body; processus terminalis about 0.9 × VIa; segm. III with 12–15 rhinaria (Quednau: 6–20) on basal ½–⅔. Apical segm. of rostrum 1.0 × 2sht. or shorter. Wings paler than in *platani;* radial sector present (Fig. 121) or absent. – Hairs of nymphs capitate, pale, placed on black tubercles. – 1.6–1.8 mm.

Oviparous female. White. Dorsum with 4 rows of small conical tubercles, provided with about 0.10 mm long, capitate hairs. Antenna about 0.7 × body. Hind tibia swollen, with many indistinct scent plaques. Cauda and anal plate rounded.

Alate male. Dorsal tubercles black, smaller than in alate viviparous female. Antenna slightly longer than body; segm. III with 20–24 secondary rhinaria, IV with 1–8, V with 3. Wings pale.

Distribution. In Sweden found in Lund in Sk. (by Danielsson); not in Denmark, Norway, or Finland. – Previously known only from C, E & S Russia, Caucasus, Kazakhstan, C Asia, Mongolia, and Korea (Quednau 1979).

Biology. The species lives on leaves of *Ulmus* spp. Danielsson collected it on *U. glabra* in Sweden on 3.IX.1978. Nevsky described it from *U. campestris (= carpinifolia).* Hosts mentioned by Shaposhnikov (1964) are *U. foliacea (= carpinifolia), pinnatoramosa, scabra,* and *suberosa.*

Note. The Swedish viviparous female specimens examined by me do not agree with the descriptions given by Nevsky (1929b) and Quednau (1979) in every respects: the wings are almost pale, without brown bordering at apical parts of veins; radial sector is far from complete, slightly visibly only at base; the middle part of pronotum is pale; and the siphunculi are not black.

Sexuales have not been found in Scandinavia; the descriptions of the oviparous female and the male are based on Nevsky's original description.

The type material from C Asia is unknown and perhaps lost. According to unpublished information from Dr. H. Szelegiewicz and Dr. F. W. Quednau more than one species have been described by various authors under the name of *saltans* Nevsky. The material described by Quednau (1979) was collected by Szelegiewicz in Korea. This species has later been published in China under the name of *yinchuanensis* Zhang, 1980. The species described by Richards (1967) under the name of *saltans* on the basis of material collected by Remaudière in Iran, is different from the species known from C Asia, as described by Nevsky, and E Asia, as described by Quednau. The chief distinguishing characters of the species from Iran are the absence of dark spots on the wings and the equal lengths of the apical segm. of rostrum and the second segm. of hind tarsus (the apical segm. of rostrum is shorter than the second segm. of hind tarsus in the material from Korea). The alatae collected by Danielsson in Sweden share these characters. Szelegiewicz (1978b) recorded *saltans* from Poland, but is now going to describe his material, which probably is conspecific with the Swedish material, as a new species in the near future.

Genus *Eucallipterus* Schouteden, 1906

Eucallipterus Schouteden, 1906: 31.
Type-species: *Aphis tiliae* Linné, 1758.
Survey: 193.

All viviparous females alate. Antenna about as long as body; processus terminalis shorter than VIa; segm. III with transverse oval, rather narrow secondary rhinaria. Antennal hairs short. Fore coxae considerably enlarged (Fig. 32). First tarsal segm. with 5–7 hairs. Siphunculi truncate. Cauda knobbed. Anal plate bilobed.

Two species in the world, one species in Scandinavia.

88. ***Eucallipterus tiliae*** (Linné, 1758)

Plate 3: 3. Figs. 32, 122–125.

Aphis tiliae Linné, 1758: 452. – Survey: 193.

Alate viviparous female. Yellow or orange; abdomen with two rows of dark dorsal spots, and also with marginal spots on some segments. Head (Fig. 122) and pronotum with dark lateral longitudinal stripes. Antennae black with middle part of segm. III (Fig. 124), bases of segm. IV, V, and VIa, and processus terminalis, pale. Distal parts of hind femora, apices of tibiae, and tarsi, dark. Abd. segm. IV with large, dark, wart-like marginal tubercles. Siphunculi usually dark. Frons with 5 small tubercles (Fig. 122). Processus terminalis 0.55–0.7 × VIa; ant. segm. III with 8–18 rhinaria placed in a row on basal half. Rostrum reaches past fore coxae; apical segm. about 0.65 × 2sht. Fore wings with dark anterior edge and dark spots at apices of oblique veins. 1.8–3.0 mm.

Oviparous female. Dorsum dark, strongly sclerotized and pigmented except in midline and along segmental borders; most body segments with broad pleurospinal cross bars and marginal sclerites. Large subsiphuncular wax gland plates present. Frons as shown in Fig. 123. Antennae without secondary rhinaria. Hind tibiae strongly swollen, rather pale, with numerous scent plaques.

Alate male. Head and thorax more pigmented than in the alate viviparous female. Antennae brownish; secondary rhinaria scattered all over the surface of the basal part of segm. III, in a single row from the apical part of III to base of VI; segm. III with 32–37 secondary rhinaria (Fig. 125), IV with 7–10, V with 6–8, VI with 2–6.

Distribution. In Denmark very common all over the country; in Sweden also common and widespread, north to Vb.; in Norway common in the south, north to NTi; in Finland common in the south, north to ObS. – Europe, W Asia, C Asia, N Africa; very common in Great Britain, N Germany, and Poland; also known from NW & W Russia. Introduced in N America and now widespread in the USA and Canada.

Biology. The species lives on the undersides of leaves of *Tilia,* often in large numbers. The sticky honeydew may drop down like a drizzle and cover all objects under the trees.

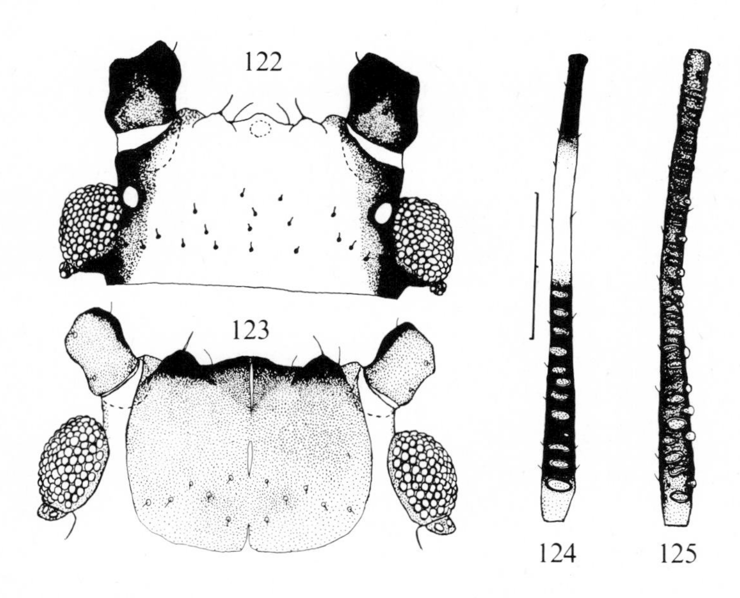

Figs. 122–125. *Eucallipterus tiliae* (L.). – 122: head of alate viviparous female; 123: head of oviparous females; 124: ant. segm. III of al. viv.; 125: ant. segm. III of alate male. (Scale 0.2 mm).

SUBTRIBE THERIOAPHIDINA

Fore coxae greatly enlarged, 1.75–3 times as broad as middle coxae (Fig. 31). – First instar nymphs with or without pleural hairs; thoracic segments with one marginal hair on each side; ant. segm. II with one hair; pronotum not fused with the head capsule.

One genus.

Genus *Therioaphis* Walker, 1870 s. lat.

Therioaphis Walker, 1870: 1999.

Type-species: *Aphis ononidis* Kaltenbach, 1846.

Survey: 423.

Yellowish, delicately built aphids with small dark dorsal spots. Apterous and alate viviparous females occur in most species, in *riehmi* only alate. Dorsal body hairs capitate, long in apterae, shorter in alatae, usually placed on wart-like spinal, marginal, and often also pleural, tubercular bases; spinal hairs of abd. tergites III, V, and VII more or less displaced laterally. Antenna as long as body or shorter, with transverse rows of fine spinules; processus terminalis from a little shorter to a little longer than VIa; segm. III with transverse oval secondary rhinaria (Fig. 131) placed in one row along segment; antennal hairs shorter than IIIbd. Rostrum not reaching middle coxae. Fore coxae greatly enlarged, more than twice as broad as middle coxae (Fig. 31). First tarsal segm. normally with 6 ventral and 2 dorsal hairs. Radial sector of fore wing obsolescent; other veins dark-bordered, ending in small dark spots. Siphunculus stump-shaped, without flange. Cauda knobbed (Fig. 133). Anal plate bilobed in viviparous females. Oviparous female without subsiphuncular wax gland plates; posterior part of abdomen prolonged and ovipositor-like.

Twenty species (two subgenera) in the world, six species in Scandinavia. The hosts are leguminous plants. The aphids are able to jump by means of muscles in the greatly enlarged fore coxae. They are not visited by ants. Sexuales usually appear in September, sometimes as early as in August. The known males are all alate.

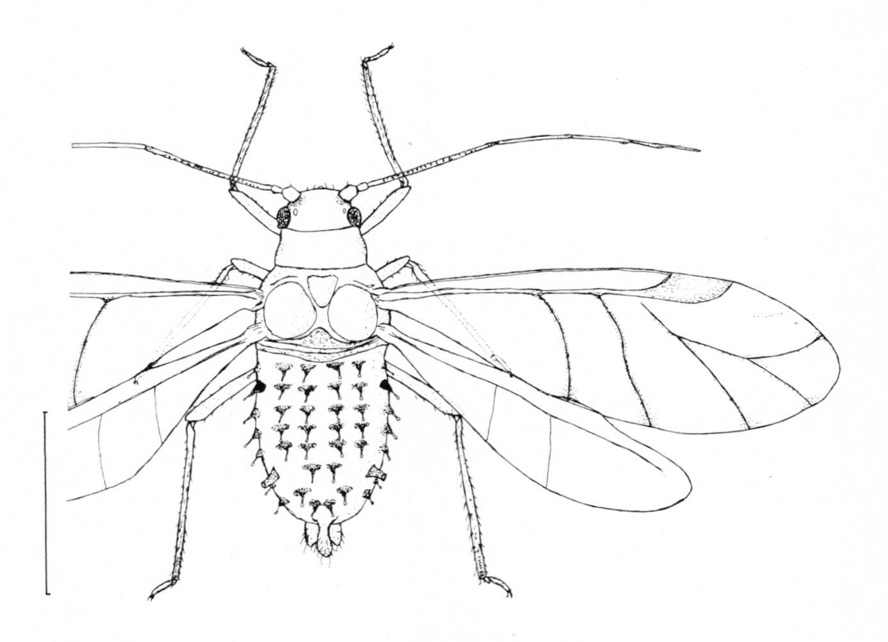

Fig. 126. *Therioaphis luteola* (Börner), al. viv. (Scale 1 mm).

Key to species of *Therioaphis*

Apterous and alate viviparous females

1 Hairs on abd. segm. I–VII not placed on wart-like tubercular bases. (Subgenus *Rhizoberlesia* del Guercio, p. 73) ... 94. *brachytricha* Hille Ris Lambers & van der Bosch
– Hairs on abdomen, or at least the marginal hairs, placed on wart-like tubercular bases. (Subgenus *Therioaphis* Walker, p. 69) ... 2
2 (1) Abd. segm. I–V each with one pair of spinal hairs and one pair of marginal hairs (Figs. 135, 137). ... 3
– Abd. segm. I–V each with more than two pairs of hairs on dorsum and margins (Figs. 127, 136, 138). ... 4
3 (2) Apical segm. of rostrum longer than 2sht. Viviparous females apterous or alate. ... 90. *ononidis* (Kaltenbach)
– Apical segm. of rostrum shorter than 2sht. All viviparous females alate. ... 91. *riehmi* (Börner)
4 (2) Pleural hairs on abd. segm. I–IV shorter than spinal hairs (Fig. 127). ... 92. *subalba* (Börner)
– Pleural hairs on abd. segm. I–IV about as long as spinal hairs. ... 5
5 (4) Abd. segm. I–IV each with one spinal pair, one pleural pair, and one marginal pair, of hairs. Alate viviparous female with secondary rhinaria on basal 82–92% of ant. segm. III. ... 89. *luteola* (Börner)
– Abd. segm. I–IV, or at least some of these, with more than six dorsal and marginal hairs. Alate viviparous female with secondary rhinaria on basal 43–71% of ant. segm. III. ... 93. *trifolii* (Monell)

Subgenus *Therioaphis* Walker, 1870 s. str.

Marginal hairs, usually also dorsal hairs, placed on wart-like tubercular bases.

89. *Therioaphis (Therioaphis) luteola* (Börner, 1949)
Figs. 126, 129, 136.

Triphyllaphis luteola Börner, 1949: 48. – Survey: 423.

Apterous viviparous female. Yellowish white or light yellow, with large, slightly pigmented, tubercular warts, each provided with a long capitate hair; abd. segm. I–V each with 5–8 such hairs, usually one spinal pair, one pleural pair, and one marginal pair; the posterior tergites with fewer hairs, tergite VIII only exceptionally with more than two spinal hairs. Marginal wart on abd. segm. II with black spot. Antenna as long as body or shorter; processus terminalis about as long as VIa; segm. III with about 10–11 rhinaria on basal two thirds, or more. Apical segm. of rostrum shorter than 2sht. 1.7–2.0 mm.

Alate viviparous female. Rather similar to the apterous viviparous female. Ant. segm. III with 14–18 rhinaria on basal 82–92% (Fig. 129).

Oviparous female. Very much like the apterous viviparous female. Ant. segm. III with about 11–13 rhinaria. Hind tibiae strongly swollen, with many scent plaques.

Alate male. Similar to the alate female. With more secondary rhinaria, on ant. segm. III about 16–18 along entire segment.

Distribution. In Denmark known from F, SZ, and NEZ; in Sweden from Sk., Öl., Ög. and Vg.; not in Norway; in Finland known from N. – The British Isles, the Netherlands, Germany, Poland, Russia, Austria, Czechoslovakia, Yugoslavia.

Biology. The host is *Trifolium pratense.*

90. ***Therioaphis (Therioaphis) ononidis*** (Kaltenbach, 1846)
Plate 3: 7. Fig. 135.

Aphis ononidis Kaltenbach, 1846: 173. – Survey: 423.

Apterous viviparous female. Yellow or orange; with dark spots, which surround large dark tubercular warts provided with long capitate hairs; abd. segm. I–VII each with one spinal pair and one marginal pair of warts; the spinal warts on tergites III, V, and VII are displaced laterally (Fig. 135); tergite VIII usually with 4 hairs. Antenna almost as long as body; processus terminalis a little longer than VIa; segm. III with 5–15 rhinaria on basal 50–60%. Apical segm. of rostrum longer than 2sht., with 10–15 accessory hairs. 1.8–2.0 mm.

Alate viviparous female. Rather similar to the apterous viviparous female. Ant. segm. III with 11–20 rhinaria on basal 47–82%.

Distribution. In Denmark rather common, known from EJ, NWJ, NEJ, and F; in Sweden known from Sk. and Öl.; not in Norway or Finland. – Europe and Asia, east to C Asia and Turkey, south to Spain; known from the British Isles, N Germany and Poland. Widespread in N America.

Biology. The species lives on *Ononis* spp. with reddish flowers *(O. repens, O. spinosa, O. hircina).*

91. ***Therioaphis (Therioaphis) riehmi*** (Börner, 1949)
Fig. 137.

Myzocallidium riehmi Börner, 1949: 49. – Survey: 423.

Alate viviparous female. Yellowish, with marginal tubercular warts, which are dark on at least abd. segm. II, IV, and V; also two longitudinal rows of dark-bordered dorsal spots or short transverse bands (Fig. 137). The marginal warts each with one hair; the dorsal spots each with one very short hair (rarely two), sometimes placed on weakly developed tubercular bases; abd. tergite VIII with only two spinal hairs. Antenna

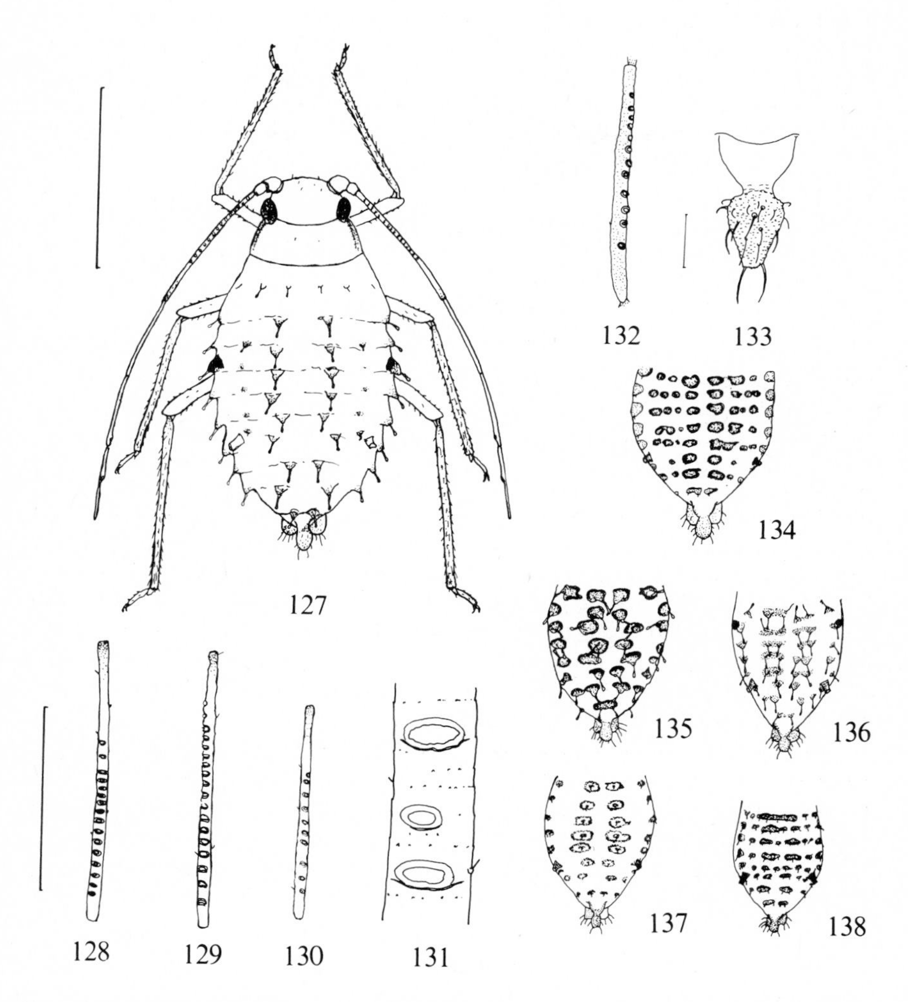

Figs. 127–138. *Therioaphis* spp. – 127: *subalba* (Börner), apt. viv.; 128–130: ant.. segm. of *subalba*, apt. viv. (128), *luteola* (Börner), al. viv. (129), and *trifolii* (Monell), al. viv. (130); 131: part of ant. segm. III of apt. viv. of *subalba* at a larger scale; 132: ant. segm. III of al. viv. of *brachytricha* H. R. L. & v. d. Bosch; 133: cauda of apt. viv. of same; 134: abdomen of apt. viv. of same; 135: abdomen of apt. viv. of *ononidis* (Kalt.); 136: abdomen of apt. viv. of *luteola;* 137: abdomen of al. viv. of *riehmi* (Börner); 138: abdomen of al. viv. of *trifolii.* (Scales 1 mm for 127, 0.5 mm for 128–130, 0.1 mm for 133). (132 and 133 after Remaudière, 134–138 after Szelegiewicz, redrawn; others original).

shorter than body; processus terminalis of about same length as VIa; segm. III with 9–12 rhinaria on basal 41–67%. Apical segm. of rostrum shorter than 2sht., with up to 6 accessory hairs. 2.0–2.7 mm.

Oviparous female. The only apterous morph. Body hairs longer than in the alate viviparous female. Ant. segm. III with about 8–9 rhinaria. Hind tibiae strongly swollen, with many scent plaques.

Alate male. With very narrow dorsal transverse bands.

Distribution. In Denmark known from EJ, NWJ, and NEZ; in Sweden found in Sk., Upl., and Vstm.; not in Norway or Finland. – In Europe east to the USSR, south to Corsica and Bulgaria; rare in Great Britain, not rare in N Germany, known from Poland. Widespread in the USA and Canada.

Biology. The aphids are found on the undersides of leaves of *Melilotus* spp., especially *M. albus*. All viviparous females are alate.

92. ***Therioaphis (Therioaphis) subalba*** Börner, 1949
Figs. 127, 128, 131.

Therioaphis subalba Börner, 1949: 49. – Survey: 423.

Apterous viviparous female. Yellowish, with six longitudinal rows of slightly pigmented, tubercular warts, one spinal pair, one pleural pair, and one marginal pair on each of the anterior abd. segments; pleural warts smaller than spinal warts on the same tergite; each wart with a capitate hair; pleural hairs shorter than spinal hairs on the same tergite, absent from some segments (Fig. 127). Marginal wart on abd. segm. II with black spot. Ant. segm. III with secondary rhinaria on basal 50–70% (Fig. 128). Otherwise very similar to *luteola.*

Alate viviparous female. Pleural hairs absent from many tergites. Ant. segm. III with secondary rhinaria on basal 60–80%.

Distribution. In Sweden occurring from Ög. and Vg. in the south to Jmt. in the north; probably occurring in southern Norway (a trapped specimen identified with some doubt according to Tambs-Lyche in litt.); in Finland known from N; not in Denmark. – C Europe; not in Great Britain and N Germany, rare in Poland (not found in the Baltic region), not in Russia.

Biology. The hosts are *Trifolium medium* and (in C Europe according to original description) *T. alpestre.*

93. ***Therioaphis (Therioaphis) trifolii*** (Monell, 1882)
Figs. 130, 138.

Callipterus trifolii Monell, 1882: 14.
Chaitophorus maculata Buckton, 1900: 277.
Survey: 424.

Apterous viviparous female. Yellow or orange, with dark spots, each carrying a low tubercular wart provided with a long capitate hair; abd. segments I–VII each with (5–)6 such dorsal hairs and also with a pair of marginal hairs; some spots on the same tergite may be fused; abd. tergite VIII with 3–6 hairs. Antenna almost as long as body; processus terminalis about as long as VIa; segm. III with 5–8 rhinaria on basal two thirds. Apical segm. of rostrum shorter than 2sht. 1.5–1.9 mm.

Alate viviparous female. Ant. segm. III with 5–14 rhinaria on basal 43–71% (Fig. 130).

Oviparous female. Very much like the apterous viviparous female. Hind tibia swollen with about 35–50 scent plaques on basal $^2/_3$.

Alate male. Antenna as long as body or longer. Secondary rhinaria on ant. segm. III: 15–19 along entire segment, IV: 3–7, V: 3–7.

Distribution. Widespread in Denmark, and in Sweden north to Jmt. and Vb.; in Norway caught in trap in AK; in Finland known from N. – Nearly all over the world, originally palaearctic. All over Europe; rare in Great Britain, not rare in N Germany, known from the Baltic coast of Poland and NW & W Russia. Asia: Middle East, Caucasus, W Siberia, C Asia, India, and China. Africa: Egypt. Introduced in N America and now widespread in the USA, the eastern part of Canada, and Mexico. Also introduced in Australia.

Biology. The species feed on several leguminous herbs, e.g. *Trifolium, Medicago, Ononis,* and *Lotus.* The Danish samples are from *Trifolium pratense* and from traps. Most of the Swedish samples are from *Trifolium (agrarium, arvense, pratense, repens),* the remainder from *Medicago falcata, M. lupulina,* and *Lotus corniculatus.*

Note. The species is in N America represented by two different strains, by several American authors (e.g. Smith & Parron 1978) still regarded as two species: *trifolii* (Monell), the Yellow Clover Aphid (YCA), and *maculata* (Buckton), the Spotted Alfalfa Aphid (SAA). Both strains are widespread in the USA, YCA also in the eastern part of Canada, SAA also in Mexico. It has been suggested that each strain is derived from a single, introduced, parthenogenetic female, or from a colony derived from a single maternal individual, because of the small range of morphological variability. The differences with regard to morphology, host plant affinities, and life cycles do not exceed those existing between populations of *trifolii* in the Old World (Dickson 1959, Ossiannilsson 1959, Hille Ris Lambers & van der Bosch 1964, Carver 1978).

Subgenus *Rhizoberlesia* del Guercio, 1915

Rhizoberlesia del Guercio, 1915: 246.

Type-species: *Rhizoberlesia trifolii* del Guercio, 1915

= *Therioaphis (Rhizoberlesia) brachytricha* Hille Ris Lambers & van der Bosch, 1964.

Survey: 373.

Dorsal hairs very short, not placed on wart-like tubercular bases except on head and abd. tergite VIII.

Two species in the world, one species in Scandinavia.

94. ***Therioaphis (Rhizoberlesia) brachytricha*** Hille Ris Lambers & van der Bosch, 1964
Figs. 132–134.

Rhizoberlesia trifolii del Guercio, 1915: 246 (sec. homonym).
Therioaphis (Rhizoberlesia) brachytricha Hille Ris Lambers & van der Bosch, 1964: 43.
Survey: 424.

Apterous viviparous female. Pale yellowish; thorax and abdomen with greyish, dark-bordered spots, viz. nearly rectangular spinal spots, smaller pleural spots, roundish, marginal spots, and small spots between the pleural spots and the marginal spots; the dorsal spots may be more or less fused (Fig. 134). Usually with more than six very short, blunt or slightly capitate, dorsal and marginal hairs on each segment. Only on the head and on abd. tergite VIII are the hairs placed on small tubercular bases. Antenna shorter than body; processus terminalis as long as VIa, or a little shorter; segm. III with 4–16 rhinaria on basal 60% or more. Apical segm. of rostrum shorter than 2sht 1.3–2.0 mm.

Alate viviparous female. Dorsal spots darker than in the apterous viviparous female. Ant. segm. III with 8–18 rhinaria on basal 75–90% (Fig. 132).

Oviparous female. Very much like the apterous viviparous female. Hind tibiae swollen.

Alate male. Similar to the alate viviparous female, but abdomen with fewer and smaller dorsal spots. Secondary rhinaria on ant. segm. III: about 15, IV: about 2, V: about 3.

Distribution. Widespread in Sweden from Sm. and Öl. in the south to Jmt. in the north; in the southern part of Norway: AK, Os, HO; not in Denmark and Finland. – Germany (not N Germany), Poland (not the Baltic coast), France, Australia, Czechoslovakia, Hungary, Yugoslavia, Turkey.

Biology. The aphids are found on the uppersides of leaves of *Lotus corniculatus* growing on dry soils.

TRIBE SALTUSAPHIDINI

Viviparous females apterous or alate; apterae usually much more common than alatae. Compound eyes protruding, without ocular tubercles. Frons convex. Antennae shorter than body, with rings of spinules. Secondary rhinaria circular or subcircular, surrounded by pale rims with dots or very short hairs; if present in apterae then only on ant. segm. III, usually on distal part of that segment. Accessory rhinaria on ant. segm. VI close together as in Callaphidini (Fig. 180) except in *Iziphya* (Fig. 212) and *Saltusaphis*. Rostrum short. Legs in some genera normal; fore and middle legs in other genera modified for leaping by means of muscles in the thickened femora. First tarsal segments with 5 hairs. Empodial hairs simple or spatulate. Cauda knobbed (Fig. 176). Anal plate bilobed, not only in viviparous females, but also in oviparae. – Head and pronotum fused in first instar nymphs.

The hosts are monocotyledones, in most cases *Carex* (Cyperaceae). The aphids are usually not visited by ants.

Key to genera of Saltusaphidini

Apterous and alate viviparous females

1 Fore and middle tibiae not constricted at base, with a smooth, sclerotic, dark "knee-cap" (Figs. 186, 213). Body oval or rather elongate. Siphunculi stump-shaped or conical, about as high as diameter of aperture or higher, placed on anterior part of margins of abd. segm. VI, apparently at the border between segm. V and VI, or just behind it (Figs. 185, 216). 2

– All tibiae constricted at base, without a sclerotic "knee-cap" (Fig. 175). Body elongate. Siphunculi formed as slightly raised pores, placed on margins of abd. segm. VI, usually near their middle (Figs. 142, 143). 4

2 (1) Accessory rhinaria on ant. segm. VI placed close to the primary rhinarium. Dorsal hairs mushroom-shaped. . *Nevskyella* Ossiannilsson (p. 98)

– Accessory rhinaria on ant. segm. VI not placed close to the primary rhinarium (Fig. 212). Dorsal hairs not mushroom-shaped. 3

3 (2) Siphunculi on level with borderline between abd. tergites V and VI. .... *Iziphya* Nevsky (p. 101)

– Anterior part of siphunculus or siphuncular cone behind borderline between abd. tergites V and VI. .... *Saltusaphis* Theobald (p. 97)

4 (1) Dorsal hairs pointed. Sometimes with visible wax pores. Apterae sometimes with secondary rhinaria. .... *Thripsaphis* Gillette (p. 75)

– Dorsal hairs mushroom-shaped (Fig. 174), at least some of them. Without wax pores. Apterae without secondary rhinaria. .... *Subsaltusaphis* Quednau (p. 87)

Genus *Thripsaphis* Gillette, 1917 s. lat.

Thripsaphis Gillette, 1917: 193.

Type-species: *Brachycolus ballii* Gillette, 1908.

Survey: 426.

Body elongate. Wax pores distinctly visible or apparently absent. Dorsal hairs all pointed. Cuticle with minute spinules (microtrichiae) or nodules. Frons convex to strongly produced medially. Rostrum short, reaching to mesosternum. Veins of fore wings often bordered with brownish.

With 18 species in the world, 8 species in Scandinavia. They live on leaves of sedges.

The genus is subdivided into four subgenera, three of which have Scandinavian representatives.

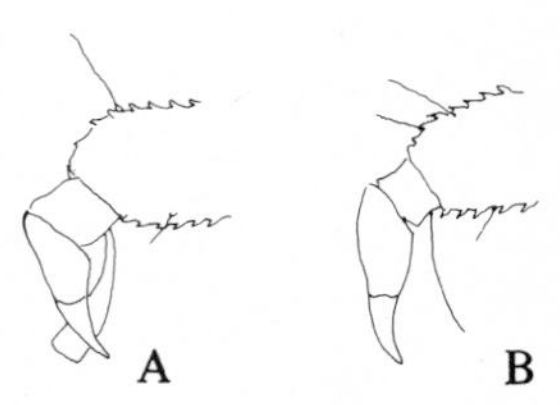

Fig. 139. Outline of apex of second tarsal segm. of *Thripsaphis* s. str. (A) with spatulate empodial hairs, and subgenus *Trichocallis* (B) with simple empodial hairs.

Key to species of ***Thripsaphis*** s. lat.

Apterous viviparous females

1 Empodial hairs spatulate (Fig. 139A). 2
– Empodial hairs simple (Fig. 139B) (subgenus *Trichocallis* Börner, p. 78) 4
2 (1) Antennae 5-segmented (subgenus *Larvaphis* Ossiannilsson, p. 86) .... 102. *brevicornis* Ossiannilsson
– Antennae 6-segmented (subgenus *Thripsaphis* Gillette, p. 77) 3
3 (2) Ant. segm. I–II dark. Antenna longer than 0.5 × body. Dorsal wax pores visible (Fig. 145). .... 96. *caricicola* (Mordvilko)
– Ant. segm. I–II pale. Antenna shorter than 0.5 × body. Dorsal wax pores invisible. .... 95. *ballii caespitosa* (Ossiannilsson)
4 (1) Posterior margin of abd. tergite VIII with a short, blunt median prominence in apterae usually hiding the cauda (Fig. 149). .... 97. *caricis* (Mordvilko)
– Posterior margin of abd. tergite VIII slightly emarginate, straight, or evenly rounded, without median prominence. 5
5 (4) Posterior margin of abd. tergite VIII slightly emarginate (Fig. 157). Wax pores apparently absent. Frons with strongly produced, rather squarish, median process (Fig. 156). .... 100. *verrucosa* Gillette
– Posterior margin of abd. tergite VIII straight or rounded (Fig. 151). Wax pores visible (Figs. 152, 160). Frons convex or with a somewhat produced, not squarish, median process. 6

6 (5) Marginal hairs on abd. tergite VIII placed in a curve rather close to posterior margin. Disc of same tergite with 6–15 shorter hairs (in the European subspecies) (Fig. 154). Ant. segm. III without rhinaria. .... 99. *ossiannilssoni* Hille Ris Lambers

– Hairs on abd. tergite VIII irregularly arranged in a curve in good distance from the posterior margin. Disc of same tergite without hairs (Fig. 153). Ant. segm. III usually with rhinaria. 7

7 (6) Flagellum (= ant. segm. III–VI) 1.4 times as long as greatest width of body or longer (Fig. 151). .... 98. *cyperi* (Walker)

– Flagellum 1.25 times as long as greatest width of body or shorter (Fig. 159). .... 101. *vibei arctica* Hille Ris Lambers

Subgenus *Thripsaphis* Gillette, 1917 s. str.

Antenna 6-segmented. Empodial hairs spatulate.

With four species in the world, two species in Scandinavia.

95. ***Thripsaphis (Thripsaphis) ballii*** (Gillette, 1908)

Brachycolus ballii Gillette, 1908: 67. – Survey: 426.

The species is subdivided into three subspecies. Two of them were described from N America, the third from Sweden.

Thripsaphis (Thripsaphis) ballii caespitosa Ossiannilsson, 1954. Figs. 140–142.

Thripsaphis caespitosa Ossiannilsson, 1954b: 118. – Survey: 426.

Apterous viviparous female. Pale dirty yellowish. Antennae, except segm. I and II and base of III, black (Fig. 141). Tarsi dark. Dorsum with membranous segmental borders, rather pale marginal sclerites, dorsal segmental plates consisting of many fused small sclerites, and a number of unsclerotized areas. Dorsal hairs irregularly arranged. Antenna 0.4 × body, without secondary rhinaria; processus terminalis 0.5–0.6 × VIa; longest antennal hair about as long as apical diameter of segm. III. Abd. tergite VIII as in *caricicola,* with hind margin medially almost straight. 1.9–2.3 mm.

Alate viviparous female. Head and pronotum with pale median longitudinal stripe. Abdomen with dark marginal sclerites; tergites III–VIII with broad, oval dorsal cross bars, segm. II with small sclerites. Antenna about 0.55 × body; segm. III with 6–8 rhinaria in a row for entire length.

Oviparous female. Greyish yellow. Much like the apterous viviparous female. Hind tibiae thickened, with numerous scent plaques.

Apterous male. Light yellow. Antenna 0.7 × body; secondary rhinaria on segm. III: 5–9 along entire segment, IV: 0, V: 0–6, mainly on distal half, VIa: 0–6.

Distribution. The subspecies has only been found in Sweden: Upl., Öl., and T. Lpm.

Biology. The subspecies lives on *Carex caespitosa.* Sexuales appear in the beginning of September in central Sweden.

Note. *T. ballii caespitosa* has less distinct microtrichiae than *b. ballii* and is paler. Ant. segm. VI of apterous *b. caespitosa* is longer than segm. III, while it usually is shorter than segm. III in *b. ballii. T. b. longisetosa* Richards, 1971, has longer hairs on body and antennae. According to the original description the longest hair on posterior margin of abd. tergite VIII is as long as 2sht., while it is shorter in *b. ballii* and *b. caespitosa.*

96. ***Thripsaphis (Thripsaphis) caricicola*** (Mordvilko, 1914)
Figs. 143–147.

Callaphis caricicola Mordvilko, 1914: 27.
Thripsaphis gelrica Hille Ris Lambers, 1956: 243.
Survey: 426.

Apterous viviparous female. Dark greyish brown, clothed with bluish grey wax powder. Antennae (except segm. III at base) and legs dark. Dorsum sclerotized, dark, with paler segmental borders and median longitudinal stripe, and darker spinal, pleural and marginal spots and intersegmental sclerites. Small wax pores present (Fig. 145). Spinal, pleural, and marginal hairs about twice as long as the irregularly placed additional hairs. Antenna about 0.6–0.7 × body; without secondary rhinaria; processus terminalis about 0.6–0.7 × VIa; longest antennal hair almost half as long as diameter of segm. IV. Apical segm. of rostrum blunt, 0.4 × 2sht. Abd. tergite VIII semilunar. 2.5–2.8 mm.
Alate viviparous female. Ant. segm. III with about 25 rhinaria.

Distribution. In the southern part of Sweden, north to Dlr.; not in Denmark, Norway, or Finland. - The Netherlands, Germany (including the northern part of Holstein, according to Gleiss (1967)), Poland (not the Baltic region), N Russia, Czechoslovakia.

Biology. The hosts are *Carex rostrata* (= *inflata*) and *vesicaria.*

Subgenus *Trichocallis* Börner, 1930

Trichocallis Börner, 1930: 127.
Type-species: *Allaphis caricis* Mordvilko, 1921
Survey: 434.

Antennae 6-segmented. Empodial hairs simple, not spatulate.
With 12 species in the world, 5 species in Scandinavia.

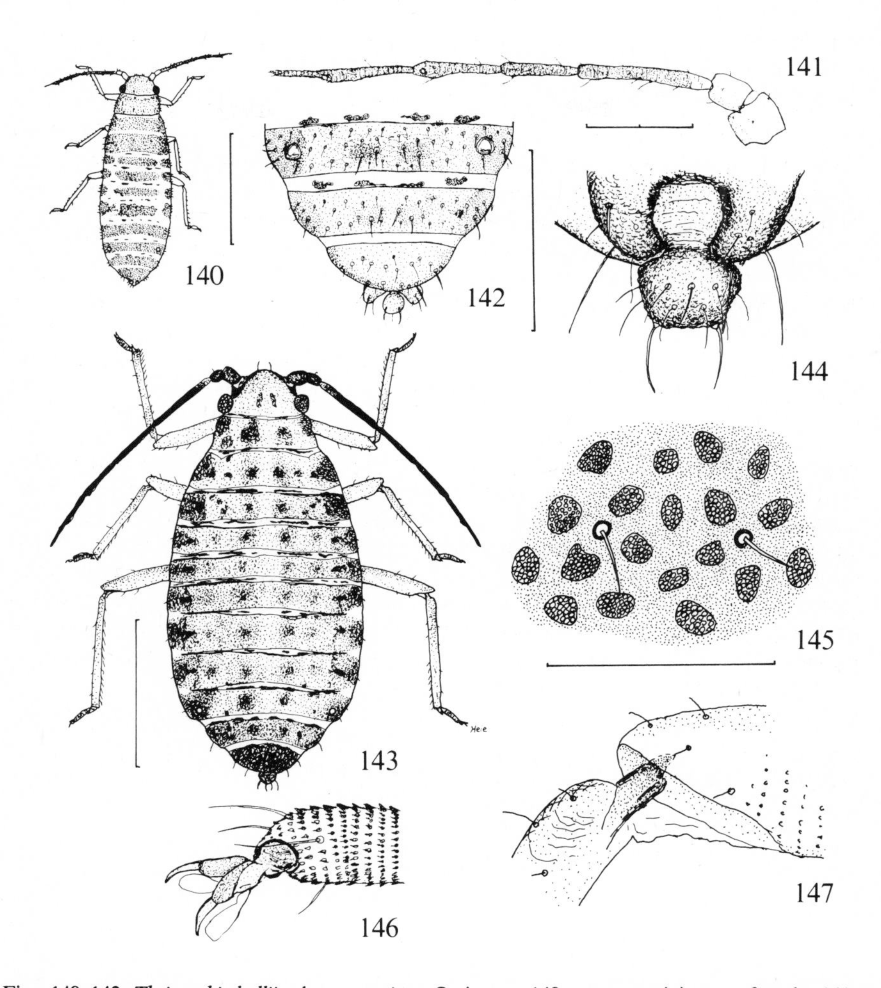

Figs. 140–142. *Thripsaphis ballii* subsp. *caespitosa* Ossiann. – 140: apterous viviparous female; 141: antenna of same; 142: posterior part of abdomen of same, dorsal view (paratype). (Scales 1 mm for 140, 0.2 mm for 141, 0.5 mm for 142). (140 after Ossiannilsson, redrawn, 141 and 142 orig.).

Figs. 143–147. *T. caricicola* (Mordv.). – 143: apterous viviparous female; 144: posterior end of abdomen, ventral view, with bilobed anal plate and cauda, below the semicircular abd. tergite VIII (seen behind); 145: part of cuticle of abd. tergite II with wax pores; 146: tip of fore tarsus with broadly spatulate empodial hairs below the claws; 147: knee of fore leg. (Scales 1 mm for 143, 0.05 mm for 145).

97. ***Thripsaphis (Trichocallis) caricis*** (Mordvilko, 1921)
Figs. 148–150.

Allaphis caricis Mordvilko, 1921: 58.
Thripsaphis thripsoides Hille Ris Lambers, 1939b: 109.
Survey: 427.

Apterous viviparous female. Yellowish white to pale greyish brown, with short tuft of wax on apex of abdomen. Dorsum sclerotized, either quite pale, or with dorsal intersegmental and marginal spots, or brownish with pale head and pale median longitudinal stripe to abd. tergite III, and dark margins as well as very dark posterior edge of tergite VIII. Wax pores present, but usually not visible except along posterior margin of abd. tergite VIII in brown specimens. Cuticle with blunt spinules or nodules arranged in irregular transverse rows (Fig. 150). Dorsal hairs very short, thorn-like. Abd. tergite VIII with 8–12 hairs. Antenna half as long as body or a little shorter; segm. III longer than segm. IV + V; processus terminalis 0.5–0.75 × VIa; segm. III with 1–5 rhinaria on distal third; antennal hairs up to about 1.0 × IIIbd. Apical segm. of rostrum short, bluntly triangular, about 0.5 × 2sht. Posterior margin of abd. tergite VIII with a short, blunt prominence with two hairs, usually covering the cauda (Fig. 149). Cauda with 9–12 hairs. 1.8–2.5 mm.

Alate viviparous female. Head and thorax black. Abdomen white or yellowish, with black marginal spots and spinopleural transverse sclerites, which are fused on segm. III–VI and here form a central shield, interrupted on I and II, very small on VII and sometimes fused with marginal sclerites on V and VI. Abd. tergite VIII dark all over. Antenna a little longer than 0.5 × body; segm. III with 10–14 rhinaria. Posterior margin of abd. tergite VIII with a projection as in the apterous viviparous female, but it does not cover the cauda.

Oviparous female. Much like the apterous viviparous female. Subsiphuncular wax gland plates large. Hind tibiae somewhat swollen, with 50–70 small scent plaques.

Apterous male. Body slender, usually constricted behind prothorax. Darker than apterous viviparous female. Antenna 0.7 × body; secondary rhinaria on segm. III: 5–10, IV: 2–5, V: 2–5; VI: 1–4. Rather small, about 1.3 mm.

Distribution. Not in Denmark; widespread in Sweden from Sk. in the south to Vb. in the north; widespread in the southern part of Norway north to SFi; widespread in Finland north to ObN. – Europe, e.g. Iceland, Great Britain (widespread, but not common), the Netherlands, N Germany (rare), and the Baltic region of Poland; south to Austria and Czechoslovakia. Records from N America, Australia, and New Zealand refer to other species (*producta* Gillette and *foxtonensis* Cottier) according to Hille Ris Lambers (1974). A subspecies, *caricis amurensis* (Mordvilko), occurs in Asia.

Biology. The species lives on leaves of *Carex* spp., in Sweden e.g. on *C. acuta* (= *gracilis*), *canescens, limosa,* and *nigra* (= *goodenowii,* sometimes spelt *goodenoughii*).

98. ***Thripsaphis (Trichocallis) cyperi*** (Walker, 1848)
Figs. 151–153.

Aphis cyperi Walker, 1848b: 45. – Survey: 427.

Apterous viviparous female. Yellowish green, covered with bluish white wax. Antennae dark. Dorsum sclerotized, rather dark, with head, segmental borders, and a median longitudinal stripe, paler. Numerous wax pores present in groups around hairs (Fig.

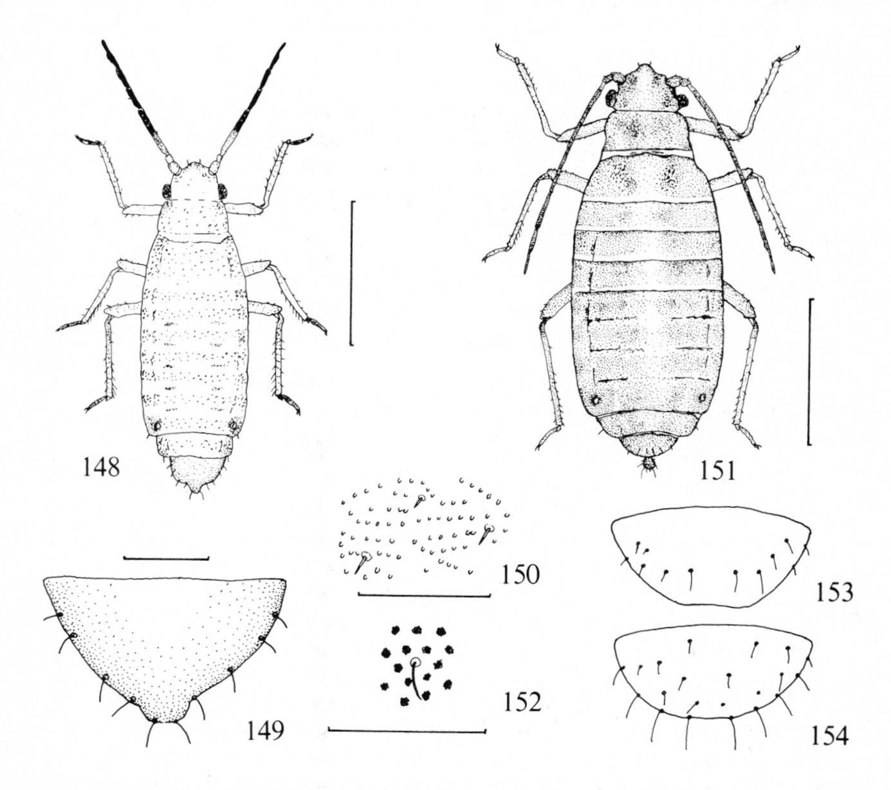

Figs. 148–150. *Thripsaphis (Trichocallis) caricis* (Mordv.). – 148: apterous viviparous female; 149: abd. tergite VIII (cauda is covered, not visible in dorsal view); 150: part of cuticle of abd. tergite II of apt. viv. (Scales 1 mm for 148, 0.1 mm for 149, 0.05 mm for 150).
Figs. 151–153. *T. (Trichocallis) cyperi* (Walker). – 151: apterous viviparous female (from Sweden, Nb.); 152: part of dorsal cuticle with wax pores; 153: arrangement of hairs on abd. tergite VIII of apt. viv. (Scales 1 mm for 151, 0.1 mm for 152).
Fig. 154. *T. (Trichocallis) ossiannilssoni* H. R. L., arrangement of hairs on abd. tergite VIII of apt. viv. (153 and 154 after Richards, redrawn; the others orig.).

152), on abd. tergites VII and VIII forming transverse bands with honeycomb-pattern. Dorsum with rather few short hairs. Abd. tergite VIII (Fig. 153) with 7–12 hairs, somewhat irregularly arranged in a curve in good distance from the posterior margin; no hairs on the disc of this tergite; distance between the two middle hairs on posterior hairs on posterior margin of abd. tergite VIII 1.25–2.0 × the distance from either of them to nearest marginal hair. Antenna about 0.65 × body; segm. III a little shorter than segm. IV + V, between 1.5 and 2 × segm. IV; processus terminalis shorter than VIa; segm. III with 0–5 rhinaria on middle, or more apically; longest hair on segm. III almost as long as IIIbd. Apical segm. of rostrum about 0.5 × 2sht. Lateral margins of abd. tergite VIII almost straight, posterior margin slightly convex, not covering cauda. 2.5–3.1 mm.

Alate viviparous female. Abdomen with transverse dorsal bands and marginal sclerites dark. Ant. segm. III with about 8–13, IV with 0–1, rhinaria.

Oviparous female. Much like the apterous viviparous female. Ant. segm. III 1.5 × segm. IV. Hind tibiae slightly swollen, with about 50 scent plaques.

Distribution. In Denmark known from WJ (Grovsø at Grærup) and EJ (Hald Sø, Rørbæk Sø); widespread in Sweden from Sk. in the south to T. Lpm. in the north; in Norway known from HOy, On, and MRy; in Finland known from Ok and ObN. – In Europe widespread, south to the Alps, east to Russia; in Great Britain widespread, more common in the north than in the south; not in N Germany; known from the southern part of Germany and from Poland, including the Baltic region, and NW, W & N Russia. N America: widespread in the USA and Canada.

Biology. The aphids live on leaves of *Carex vesicaria* and *C. rostrata*.

Note. The species is very similar to *caricicola,* but has simple empodial hairs. *T. caricicola* has the same arrangement of marginal hairs on abd. tergite VIII, but some extra hairs are present on the disc of this tergite (Hille Ris Lambers 1974), and the empodial hairs are spatulate.

99. ***Thripsaphis (Trichocallis) ossiannilssoni*** Hille Ris Lambers, 1952.
Figs. 154, 155.

Thripsaphis ossiannilssoni Hille Ris Lambers, 1952b: 56. – Survey: 427.

Apterous viviparous female. Pale, with smoky or rather dark marginal sclerites and dark abd. tergite VIII. Antennae dark. Wax pores visible around some of the hairs and on marginal sclerites of abd. tergite VII and most of tergite VIII. Dorsum with very small spinules on pigmented areas and very short hairs; abd. tergite VIII (Fig. 154) with 6–15 inconspicuous hairs on disc and a single or double row of much larger marginal hairs; distance between the two middle hairs on posterior margin of this tergite about 0.9–1.3 × the distance from either of them to nearest marginal hair. Antenna about 0.4 × body, without secondary rhinaria; segm. III considerably shorter than segm. VI and also much shorter than segm. IV + V; processus terminalis a little shorter than VIa; longest antennal hair shorter than IIIbd. Apical segm. of rostrum about 0.5–0.6 × 2sht.

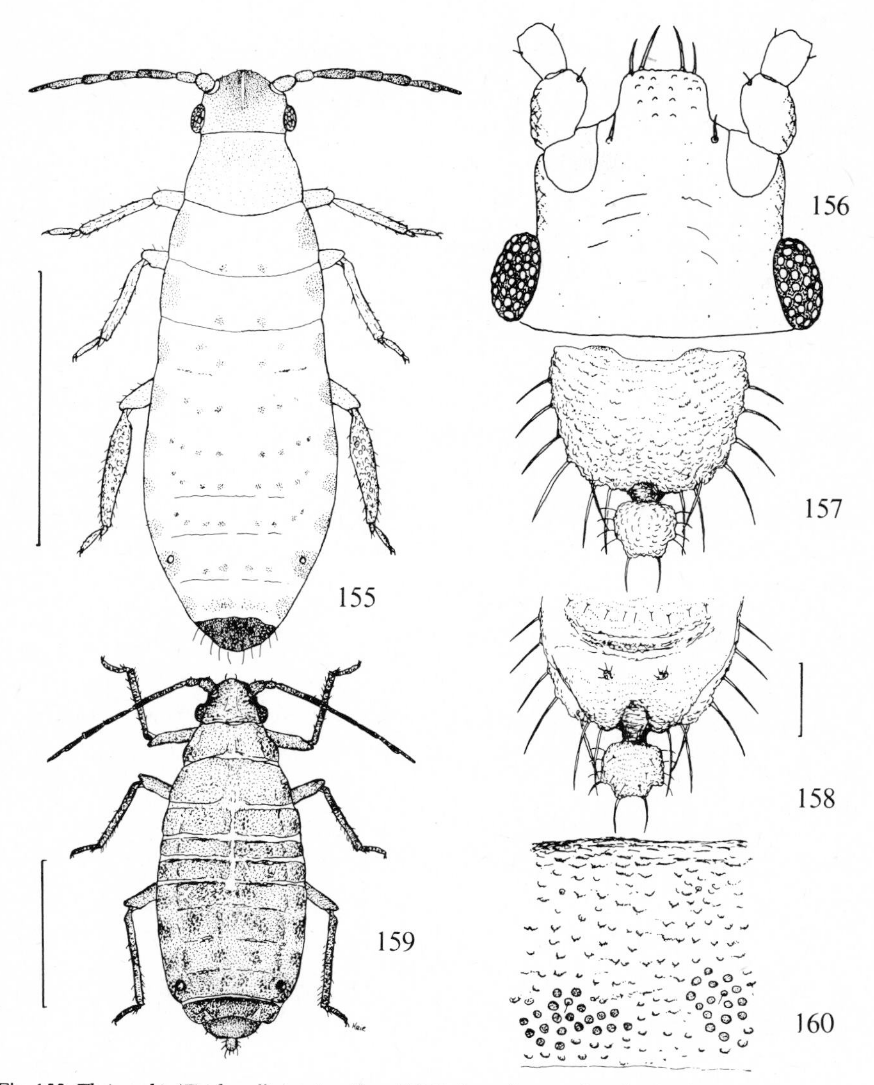

Fig. 155. *Thripsaphis (Trichocallis) ossiannilssoni* H. R. L., oviparous female. (Scale 1 mm).
Figs. 156–158. *T. (Trichocallis) verrucosa* Gill., apt. viv. – 156: head; 157: posterior part of abdomen, dorsal view, with cauda and emarginate abd. tergite VIII; 158: the same part in ventral view, with cauda, bilobed anal plate, and rudimentary gonapophyses. (Scale 0.1 mm for 156–158).
Figs. 159, 160. *T. (Trichocallis) vibei* subsp. *arctica* H. R. L., apt. viv. (from Sweden, Lu.Lpm). – 159: body; 160: middle part of abd. tergite VII with wax pores on posterior half. (Scale 1 mm for 159).

Abd. tergite VIII evenly rounded, more or less semicircular. About 1.9 mm.

Oviparous female. Much like apterous viviparous female. Subsiphuncular wax gland plates pale. Ant. segm. III 1.4 × segm. IV. Hind tibia distinctly swollen, rather pale, with paler scent plaques on most of its surface. About 2.1 mm (Fig. 155).

Distribution. In Sweden known from Öl., Sdm., and Upl.; not in Denmark, Norway, and Finland. – Described from Sweden and also recorded from the Netherlands, Germany (Berlin and Sylt), Poland, Czechoslovakia, and Japan. The subspecies *ossiannilssoni pacifica* Hille Ris Lambers, 1974, in which disc of abd. terg. VIII is free of hairs, occurs in California.

Biology. The species lives on *Carex* spp. *(C. appropinquata (= paradoxa), elata, lasiocarpa, nigra).*

100. ***Thripsaphis (Trichocallis) verrucosa*** Gillette, 1917
Figs. 156–158.

Thripsaphis verrucosa Gillette, 1917: 194. – Survey: 427.

Apterous viviparous female. Yellow or pale greenish yellow. Antenna with distal part dark, and at least segm. I and II and basal part of III, pale. Wax pores apparently absent. Dorsum sclerotized, uniformly pale, with numerous nodules and rather few very short hairs, almost hairless; abd. tergite VIII with 6–10 stout, stiff, marginal hairs. Frons with strongly produced, rather squarish median process (Fig. 156). Antenna half as long as body or longer; segm. III a little shorter than segm. IV + V, about as long as segm. VI, sometimes longer; processus terminalis 0.8–1.1 × VIa; segm. III with 0–5 rhinaria on distal half; longest hair on segm. III shorter than 0.5 × IIIbd. Apical segm. of rostrum about 0.4–0.5 × 2sht. Posterior margin of abd. tergite VIII slightly emerginate, not covering the cauda (Fig. 157). Cauda with 15–21 hairs. 2.4–2.8 mm.

Alate viviparous female. Head and thorax dark; abdomen with dark sclerites, on tergites III–VI with dark transverse bands at least partly separated at segmental borders. Antennae black. Legs dusky. Dorsal hairs very fine and inconspicuous, almost invisible. Sclerotic areas with oval nodules; on the sclerotic bands of abdomen transverse oval nodules bearing small spinules on posterior half. Antenna 0.75 × body or longer; segm. III with 9–20 secondary rhinaria.

Oviparous female. Pale greenish yellow. Much like the apterous viviparous female. Hind tibiae slightly swollen, with numerous small scent plaques.

Distribution. In Denmark found in NEJ; in Sweden widespread from Sk. in the south to T. Lpm. in the north; in Norway known from HOy; in Finland from Ok. – Widespread in Europe, e.g. the Faroes, Great Britain, and Poland, including the Baltic region; not in N Germany. Asia: W Siberia. N America: widespread in the USA and Canada.

Biology. The species lives on several *Carex* spp., in Scandinavia especially on *C.*

nigra (= goodenowii). Sexuales appear at the end of August and in September in central Sweden (Ossiannilsson 1959).

101. ***Thripsaphis (Trichocallis) vibei*** Hille Ris Lambers, 1952 s. lat.

Thripsaphis vibei Hille Ris Lambers, 1952: 25. – Survey: 427.

The species is subdivided into two subspecies, viz. *v. vibei* from Greenland (Frederikshåb and Unartoq) and *v. arctica* from Greenland (Narssarssuaq), Iceland, and N Scandinavia.

Thripsaphis (Trichocallis) vibei arctica Hille Ris Lambers, 1955.
Figs. 159, 160.

Thripsaphis cyperi arctica Hille Ris Lambers, 1955: 19.
Thripsaphis vibei arctica Hille Ris Lambers, 1960: 15.
Survey: 427.

Apterous viviparous female. Rather similar to *cyperi.* Dorsum mottled, dark, with paler median longitudinal stripe from mesonotum to anterior margin of abd. tergite III; tergites III–VI fused, other tergites free. Wax pores present in groups around spinal, pleural, and marginal hairs, on the middle part of the posterior half of tergite VII (Fig. 160), and in groups on tergite VIII. Antenna 0.55–0.65 × body; segm. III a little shorter than segm. IV + V, 1.6–1.7 × IV; processus terminalis 0.8–0.95 × VIa; segm. III with 0–3 rhinaria on distal half. Posterior margin of abd. tergite VIII straight. 2.3–2.7 mm.

Alate viviparous female. Head and thorax black. Abdomen with dark marginal sclerites and transverse dorsal bands, those on tergites III–VI broad and partly fused. Ant. segm. III with about 9–10 rhinaria.

Oviparous female. Smaller and more slender than apterous viviparous female. Abd. tergites III–VI rarely completely fused. Ant. segm. III with 0–1 rhinaria. Subsiphuncular wax gland plates very large. Basal half, or slightly more, of hind tibia indistinctly swollen, and with about 30–50 scent plaques.

Apterous male. Sclerotization as in oviparous female. Antenna about 0.65 × body; secondary rhinaria small, on segm. III: 8–10 along one side, IV: 3–5, V: 7, VIa: 3–4. About 1.8 mm.

Distribution. In Sweden found in Lu.Lpm.; not in Denmark, Norway, and Finland. – Iceland, Greenland.

Biology. The aphids live on *Carex* sp. Sexuales are produced at the end of July in Greenland, in early August in Iceland.

Note. *T. v. arctica* is intermediate between *cyperi* and *v. vibei* with regard to some characters. The body is rather short and broad in *v. vibei,* but elongate in *cyperi* and *v. arctica.* The antennae of apterous viviparous females of *v. vibei* and *v. arctica* are much shorter than in *cyperi.* The flagellum (= ant. segm. III–VI) is 0.9–1.0 × the greatest width of body in *v. vibei,* 1.15–1.25 in *v. arctica,* and 1.4–1.8 in *cyperi* (Hille Ris Lambers 1974).

Subgenus *Larvaphis* Ossiannilsson, 1953

Larvaphis Ossiannilsson, 1953: 237.
Type-species: *Thripsaphis (Larvaphis) brevicornis* Ossiannilsson, 1953.
Survey: 241.

Antenna 5-segmented. Empodial hairs spatulate.
Only one species in the subgenus.

102. ***Thripsaphis (Larvaphis) brevicornis*** Ossiannilsson, 1953
Fig. 161.

Thripsaphis (Larvaphis) brevicornis Ossiannilsson, 1953: 237. – Survey: 427.

Apterous viviparous female. Body slender, almost parallelsided. Head narrower than body. Dorsum of head, pronotum, mesonotum, and abd. tergite VIII sclerotic; other body segments with many more or less fused scleroites; segmental borders distinct. Dorsal hairs short, those on abd. tergite VIII somewhat longer, numerous on sclerotic parts, one on each scleroite of the other segments. Antenna about 0.25 × body, without secondary rhinaria; processus terminalis about 0.4 × Va, 0.25–0.30 × segm. III; longest antennal hair a little longer than IIIbd. Apical segm. of rostrum obtuse. Legs short, with numerous hairs about as long as antennal hairs. Abd. tergite VIII semicircular. 2.7–3.0 mm.

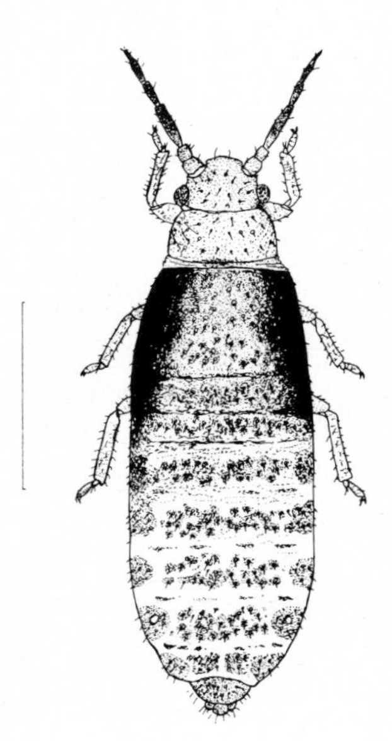

Fig. 161. *Thripsaphis (Larvaphis) brevicornis* Ossiannilsson, apt. viv. (Scale 1 mm).

Oviparous female. Much like the apterous viviparous female. Yellow, pigmented parts brownish, sparsely covered with wax. Hind tibiae swollen, with numerous scent plaques. About 2.1 mm.

Apterous male. Yellow, with muscle scleroites, and distal part of antennae, darker. Antenna 5-segmented, about 0.5 × body; processus terminalis 0.3 × Va; secondary rhinaria on segm. III: 8–14, IV: 3–7, Va: 3–5. 1.5–1.6 mm.

Distribution. In Sweden known from Upl. and Dlr.; not in Denmark, Norway, or Finland. – Also recorded from Germany (Berlin) (Quednau 1954). Mamontova (1963) has described subsp. *carpatica* from E Europe.

Biology. The aphids live on *Carex lasiocarpa* and *C. elata.* Sexuales appear at the end of September and in October in central Sweden.

Genus *Subsaltusaphis* Quednau, 1953

Subsaltusaphis Quednau, 1953: 224.

Type-species: *Saltusaphis intermedia* Hille Ris Lambers, 1939.

Survey: 413.

Body elongate, flat. Apterae with pale sclerotic dorsum, sometimes with dusky or dark lines or spots besides dark intersegmental lines or rows of muscle sclerites, without membranous borders between abd. tergites III–VI. Alatae with a central shield formed by fused cross bars on abd. tergites III–V, and smaller dorsal and marginal sclerites. Wax pores absent. Dorsal and marginal hairs stellate, i.e. short and mushroom-shaped, resembling everted umbrellas with marginal incisures (Fig. 174); normal hairs, which may be pointed, blunt, capitate, or bifurcated, are present on frons, abd. tergite VII and VIII, sometimes also on margins of other body segments, rarely on dorsum and then mixed with stellate hairs (Fig. 171). Cuticle of dorsum with numerous small nodules or denticles. Frons convex. Apterae without secondary rhinaria, alatae with secondary rhinaria on ant. segm. III. Rostrum short, reaching just past fore coxae; apical segm. very short and blunt. Legs not modified for leaping. Empodial hairs spatulate. Veins of fore wing sometimes bordered with pale brownish shadows; hind wing with only one oblique vein. Siphunculi with slightly raised pores placed at margins of abd. segm. VI. Abd. tergite VIII apically more or less incised (Figs. 162–169).

There are 14 species in the world, 8 species in Scandinavia. Two species occurring in NW Europe, but not yet found in Scandinavia, are included in the key and descriptions below.

Some species have been collected only a few times. Several of them are little known. Key and descriptions are partly based on the original descriptions and other literature (Hille Ris Lambers 1939 and 1956, Ossiannilsson 1959, Quednau 1953 and 1954, Richards 1971, Stroyan 1977, and Theobald 1929).

Key to species of *Subsaltusaphis*

Apterous viviparous females

1 Short spiny normal hairs occur on dorsum, mixed with mushroom-shaped hairs (Fig. 171). 2

– Dorsum exclusively with mushroom-shaped hairs. Normal hairs only present on frons and margins of posterior abd. segments. 3

2 (1) Mushroom-shaped hairs more numerous than normal hairs, especially on thorax and marginopleural zones of abdomen; spiny hairs present especially along median line of abdomen. 103. *aquatilis* (Ossiannilsson)

– Mushroom-shaped hairs not more numerous than normal hairs. .... *intermedia* (Hille Ris Lambers)

3 (1) Nodules rather large and flat, shaped as oval rings (Fig. 184). Incisure of posterior margin of abd. tergite VIII deeper than incisure of anal plate (Fig. 166). Abd. tergite VIII with 12–16 marginal hairs. .... 107. *pallida* (Hille Ris Lambers)

– Nodules or denticles small, shaped as points or – in a lateral view – as small blunt or pointed spinules (Fig. 182). Posterior margin of abd. tergite VIII not deeply incised, or – if so, then abd. tergite VIII with 8–10 marginal hairs. 4

4 (3) Basal parts of ant. segm. IV and V pale. 5

– Ant. segm. IV and V dark all over (but segm. IV often somewhat paler at base than at apex). 6

5 (4) Antenna longer than 0.6 × body; segm. III 1.2–1.4 × segm. VI. Margins of body pale. Dorsum without longitudinal rows of dusky spots. .... *maritima* (Hille Ris Lambers)

– Antenna shorter than 0.6 × body; segm. III 0.9–1.3 × segm. VI. Margins of body not conspicuously pale. Dorsum at least with traces of longitudinal rows of dusky spots. .... 108. *paniceae* (Quednau)

6 (4) Antenna longer than 0.75 × body. .... 105. *lambersi* (Quednau)

– Antenna shorter than 0.75 × body. 7

7 (6) Dorsum without any trace of dusky or dark longitudinal lines or longitudinal rows of pigmented spots. Abd. tergite VIII with 16–20 long marginal hairs (Fig. 163). .... 104. *flava* (Hille Ris Lambers)

– Dorsum with at least traces of dusky longitudinal lines, or with longitudinal rows of pigmented spots. Abd. tergite VIII with 6–16 long marginal hairs. 8

8 (7) Dorsum rather evenly and densely adorned with rather coarse nodules. .... 110. *rossneri* (Börner)

– Dorsum adorned with nodules, which are rather coarse in the pleural zone, but become finer and sparser towards margins and spinal zone. 9

9 (8) Pleural longitudinal dark lines or rows of spots well-marked and clearly defined on spinal side, nearly or quite as dark as the neighbouring intersegmental muscle sclerites. Mid-dorsum between dark lines strikingly pale, with at most traces of dusky sclerites in the middle of one or more of the abd. segments I–IV. .... 106. *ornata* (Theobald)

– Pleural longitudinal lines or rows of spots very ill-defined, much paler than intersegmental muscle sclerites. Mid-dorsum not conspicuously pale, usually with an ill-defined dark median longitudinal line on abd. segm. I–IV. .... 109. *picta* (Hille Ris Lambers)

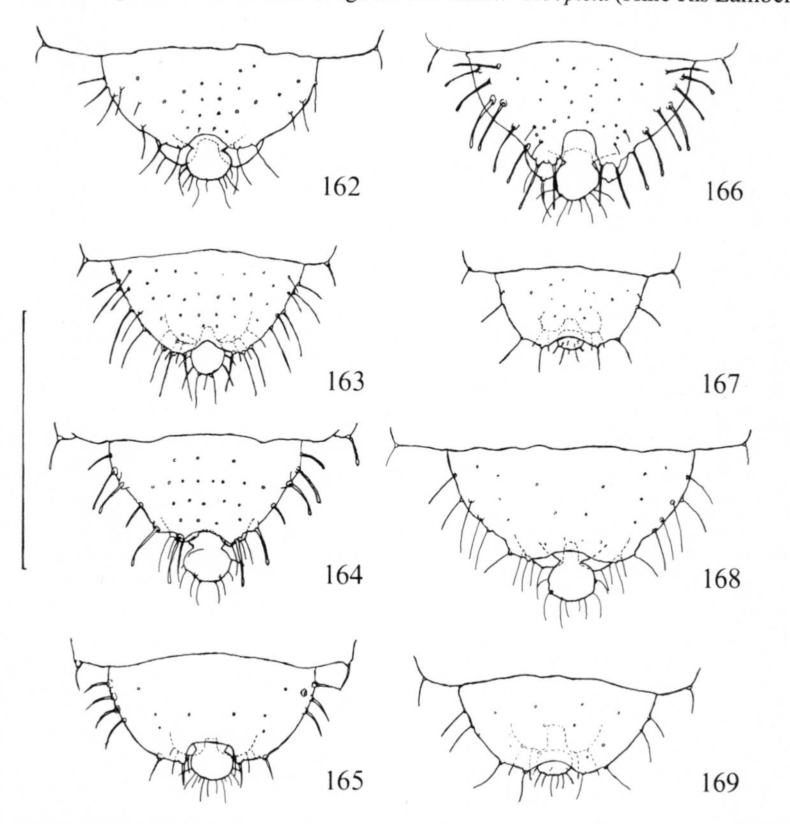

Figs. 162–169. *Subsaltusaphis* spp., apt. viv., posterior part of abdomen, dorsal view. – 162: *aquatilis* (Ossiann.); 163: *flava* (H. R. L.); 164: *lambersi* (Quednau); 165: *ornata* (Theob.); 166: *pallida* (H. R. L.); 167: *paniceae* (Quednau); 168: *picta* (H. R. L.); 169: *rossneri* (Börner). (Scale 1 mm).

103. ***Subsaltusaphis aquatilis*** (Ossiannilsson, 1959)
Figs. 162, 170, 171.

Bacillaphis aquatilis Ossiannilsson, 1959: 5. – Survey: 413.

Apterous viviparous female. Pale yellowish, with dark muscle sclerites. Antennae brown with segm. I–II, and basal half of III, pale. Tarsi dark. Short pointed hairs or small spines mixed with mushroom-shaped hairs occur along midline of dorsum and on margins of abd. segm. (II–) III–VIII (Fig. 171); they are of increasing length towards cauda. Abd. tergite VIII with about 10 marginal hairs (Fig. 162). Frons moderately convex. Antenna 0.5 × body; segm. III 1.1–1.3 × segm. VI; processus terminalis 0.5–0.8 × VIa, longest antennal hairs about 0.8 × IIIbd. Posterior margin of abd. tergite VIII with small incisure. 2.1–2.6 mm.

Distribution. Known only from Sweden: Dlr., Med., and Lu.Lpm.

Biology. The host is *Carex aquatilis*.

104. ***Subsaltusaphis flava*** (Hille Ris Lambers, 1939)
Plate 4: 4. Figs. 163, 172–176.

Saltusaphis flava Hille Ris Lambers, 1939b: 102. – Survey: 413.

Apterous viviparous female. Dull yellow, without dark markings except for narrow transverse intersegmental stripes, or rows of muscle sclerites. Antennae black with segm. I–II, and basal half of III, pale. Dorsum with numerous mushroom-shaped hairs (Figs. 173, 174). Abd. tergite VII with one pair of long marginal hairs, tergite VIII (Fig. 163) with 16–20 such hairs. Antenna 0.5 × body; segm. III about 1.0 × segm. VI; processus terminalis about 0.75–1.0 × VIa; longest antennal hairs slightly longer than 0.5 × IIIbd. Abd. tergite VII trapezoid, with rounded hindcorners. Posterior margin of abd. tergite VIII with small incisure. 1.8–2.1 mm.

Alate viviparous female. Head, thorax, and antennae dark, with paler median stripe on head and pronotum. Abdomen with the same type of sclerotization as in *ornata*. Ant. segm. III with 8–9 secondary rhinaria. Radial sector of fore wing indistinct.

Oviparous female. Rather similar to the apterous viviparous female. Hind tibia somewhat swollen on basal two thirds, with about 60 roundish scent plaques, some of which are fused two by two.

Distribution. In Denmark found in WJ, NWJ, and NEJ; in Sweden common and widespread, from Hall., Bl., Sm., and Öl. in the south to Nb. in the north; in Norway known from HOy and On; in Finland known from ObN. – British Isles, the Netherlands, Germany, Poland, Czechoslovakia, and Austria; the species is rare in N Germany and Great Britain.

Biology. The species lives on *Carex nigra (= goodenowii)*. In Denmark oviparous females occur in October.

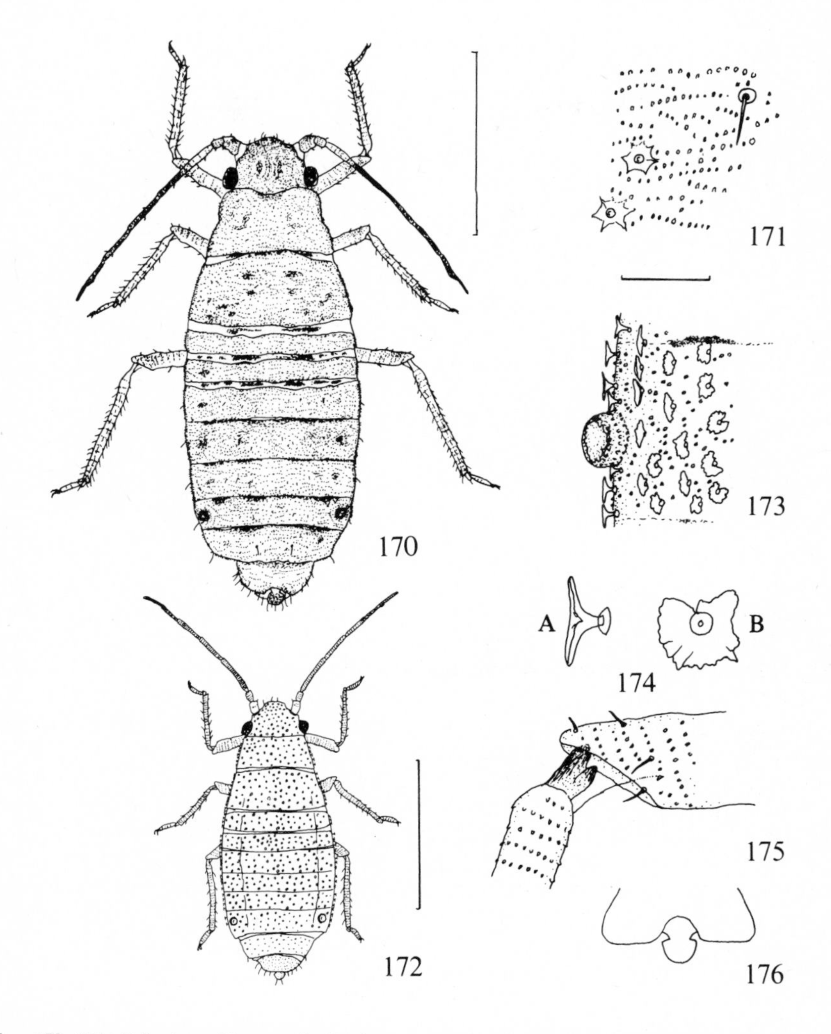

Figs. 170, 171. *Subsaltusaphis aquatilis* (Ossiann.), apt. viv. (paratype). – 170: habitus; 171: part of mid-dorsum, abd. tergite V. (Scales 1 mm for 170, 0.01 mm for 171).
Figs. 172–176. *S. flava* (H. R. L.), apt. viv. – 172: habitus; 173: left margin of abd. segm. VI with siphunculus, mushroom-shaped hairs, and nodules; 174: mushroom-shaped hair, lateral view (A) and dorsal view (B); 175: knee of fore leg; 176: anal plate and cauda in outline, ventral view. (Scale 1 mm for 172).

Subsaltusaphis intermedia (Hille Ris Lambers, 1939)

Saltusaphis intermedia Hille Ris Lambers, 1939b: 103. – Survey: 413.

Apterous viviparous female. Yellowish white, with indistinct dark intersegmental transverse rows of spinopleural and marginopleural muscle sclerites; the marginopleural pair absent from borderline between meso- and metathorax, and between abd. segm. VII and VIII. Antennae black with segm. I–II, and basal $^2/_3$–$^3/_4$ of III, pale. Dorsum with irregular transverse rows of denticles. Mushroom-shaped hairs and normal, very short, spiny hairs present on dorsum in about equal number. Margins of abd. segm. II–VII with 1–2 pairs of short normal hairs per segment, increasing in length towards the rear. Abd. segm. VII with one pair of longer marginal hairs, VIII with 6–10 still longer marginal hairs. Antenna about 0.5 × body; segm. III about 1.1 × segm. VI; processus terminalis 0.6–0.7 × VIa; antennal hairs about 0.6 × IIIbd. Posterior corners of abd. tergite VII not angular. Posterior margin of tergite VIII only slightly incised. 2.0–2.2 mm.

Alate viviparous female. Head and thorax dark. Abdomen with the same type of sclerotization as in *ornata.* Ant. segm. III with about 9–12 secondary rhinaria.

Distribution. Not yet found in Scandinavia. – The Netherlands, Germany, Austria, Czechoslovakia.

Biology. The host plant is *Carex hirta.* The aphids live between the bases of the leaves.

105. ***Subsaltusaphis lambersi*** (Quednau, 1954)
Figs. 164, 177–180.

Bacillaphis lambersi Quednau, 1954: 39, 48. – Survey: 413.

Apterous viviparous female. Pale, with narrow brownish intersegmental transverse stripes, dark margins, of thorax at least, and two brownish pleural longitudinal stripes from head to abd. tergite VIII, or a few brownish, rather pale and small pleural areas on head and thorax. Antennae dark with segm. I–II, and base of III, pale. Coarse nodules present on thorax and along pleural lines of abdomen, rather few finer nodules on other parts of body (Figs. 178, 179). Abd. tergites VI and VII with a blunt or weakly capitate hair on each posterior corner. Abd. tergite VIII (Fig. 164) with about 14 normal marginal hairs. Antenna rather long, longer than 0.75 × body, reaching as far as beyond siphunculi; segm. III about 1.3 × segm. VI; processus terminalis about 1.0 × VIa. Posterior margin of abd. tergite VIII incised. About 2.8 mm.

Distribution. Known only from Sweden: Sk., Nrk., and Upl.

Biology. The host is *Carex acuta (= gracilis).* The aphids apparently prefer shady localities (Ossiannilsson 1959). In C Sweden oviparous females occur from early September.

Subsaltusaphis maritima (Hille Ris Lambers, 1956)

Bacillaphis maritima Hille Ris Lambers, 1956: 244. – Survey: 413.

Apterous viviparous female. Yellowish white, colourless at margins, especially on posterior half of body, with pale brownish yellow transverse intersegmental lines and very small pleural spots. Antennae pale with apex of segm. III, distal $^1/_3$–$^2/_3$ of IV, most of V, and entire segm. VI, blackish. Dorsum covered with wavy transverse rows of very small denticles and numerous mushroom-shaped hairs. Blunt, or capitate, or bifurcated, stout and normal, hairs are present on abd. tergites VI–VIII. Abd. tergite VIII

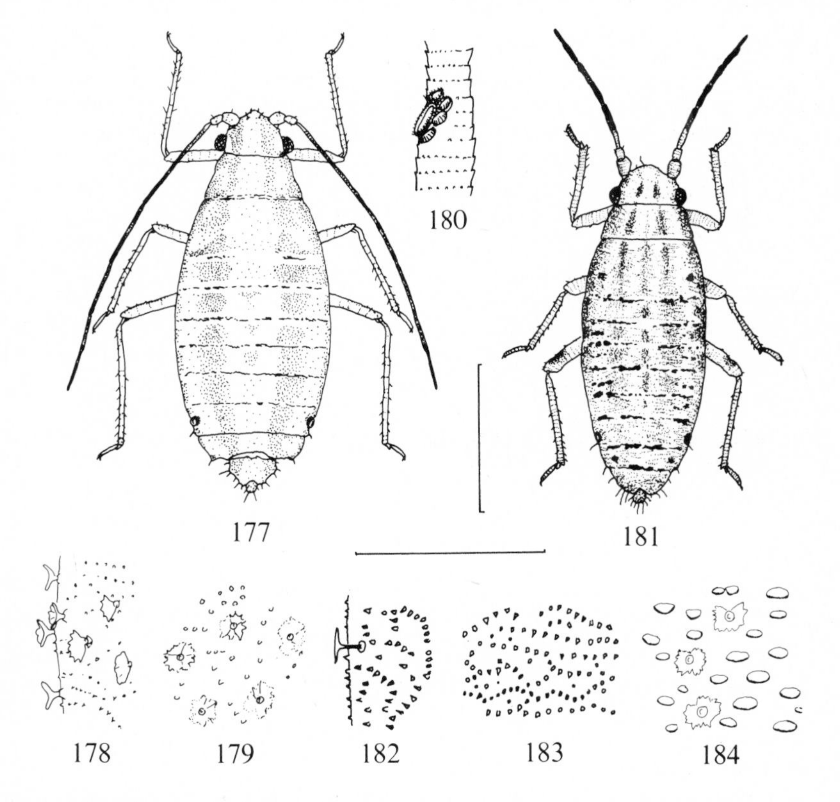

Figs. 177–180. *Subsaltusaphis lambersi* (Quednau), apt. viv. – 177: habitus; 178: part of margin of abd. segm. VII; 179: part of mid-dorsum of abd. segm. V; 180: part of ant. segm. VI with primary rhinarium and accessory rhinaria. (Scale 1 mm for 177).

Figs. 181–183. *S. rossneri* (Börner), apt. viv. – 181: habitus; 182: part of margin of abd. segm. VII; 183: part of mid-dorsum of abd. segm. V.

Fig. 184. *S. pallida* (H. R. L.), part of mid-dorsum of abd. segm. V of apt. viv. (Scale 0.05 mm).

with 8–10 hairs. Antenna 0.6–0.7 × body; segm. III 1.2–1.4 × segm. VI; processus terminalis 0.8–0.9 × VIa. Posterior margin of abd. tergite VIII rather deeply incised. 2.0–2.6 mm.

Distribution. Not yet known from Scandinavia. – The Netherlands.

Biology. The host is *Scirpus maritimus.*

106. ***Subsaltusaphis ornata*** (Theobald, 1927)
Fig. 165.

Saltusaphis ornata Theobald, 1927: 30. – Survey: 413.

Apterous viviparous female. Dorsum with two very distinct pleural longitudinal lines or rows of segmental spots, nearly as dark, or quite as dark, as the transverse intersegmental stripes or rows of muscle sclerites; without dark median longitudinal line, or at most with traces of dusky sclerites in the middle of one or more of the anterior four abd. tergites; the mid-dorsum between the pleural lines paler than the lateral parts of dorsum; margins of body rather dark. Antennae black with segm. I–II, and basal half of III, pale. Tarsi brownish. Dorsal cuticle with numerous densely placed small nodules, most distinct and coarse on pigmented areas, finer and sparser towards margins as well as towards midline of dorsum. With rather few mushroom-shaped hairs on pale parts of body. Abd. tergites VII–VIII with normal hairs. Antenna about 0.65 × body; segm. III about 1.0–1.3 × segm. VI; processus terminalis about 0.8–1.0 × VIa; longest hair on antennae about 0.9–1.0 × IIIbd. Posterior angle of abd. tergite VII about 100° or slightly more.Posterior margin of tergite VIII distinctly incised (Fig. 165). 2.7–2.8 mm.

Alate viviparous female. Head and pronotum orange, each with two short dark longitudinal stripes or spots. Abdomen pale green or yellow with rather dark marginal sclerites, those on segm. VI surrounding the siphunculi, and dorsal sclerites or transverse bands, those on tergites III–V more or less fused. Antennae dark with segm. I and II paler; segm. III with about 13 rhinaria for entire length.

Distribution. In Sweden found in Sk. (Danielsson in litt.); not in Denmark, Norway, or Finland. – Great Britain, the Netherlands, Germany, Austria, Czechoslovakia, France.

Biology. The species lives on *Carex riparia,* possibly also *C. acutiformis.*

107. ***Subsaltusaphis pallida*** (Hille Ris Lambers, 1939)
Figs. 166, 184.

Saltusaphis pallida Hille Ris Lambers, 1939b: 105. – Survey: 413.

Apterous viviparous female. Pale yellow, with dark transverse rows of intersegmental muscle sclerites. Antennae pale, with apex of segm. III, distal half of IV, and entire V and VI, black. Nodules (Fig. 184) rather large, transverse oval, much flatter than in other *S.* spp., especially on disc of abdomen. Dorsum with mushroom-shaped hairs. Abd. tergite VI with one marginal pair of thick, sometimes bifurcated, normal hairs. Abd. tergite VII with longer hairs on posterior corners. Abd. tergite VIII with 14–16 normal hairs along margins. Antenna about 0.65 × body; segm. III about 1.5 × segm. VI; processus terminalis about 0.8 × VIa. Posterior margin of abd. tergite VIII with deep incisure, deeper than incisure of anal plate (Fig. 166). About 2.6–2.7 mm.

Distribution: In Denmark known from NWJ (Vandet Sø, Thy); in Sweden known from Öl., Gtl., and Upl.; not in Norway or Finland. – The Netherlands, Austria, Czechoslovakia, Hungary, Spain, and Crimea.

Biology. The species lives on *Carex* sp.

108. ***Subsaltusaphis paniceae*** (Quednau, 1954)
Fig. 167.

Bacillaphis paniceae Quednau, 1954: 39, 48. – Survey: 413.

Apterous viviparous female. Colour pattern more or less as in *picta.* Antennae pale with distal halves of segments III–V, and segm. VI dark. Nodules and mushroom-shaped hairs as in *picta.* Normal hairs usually restricted to margins of abd. tergites VII–VIII, more rarely also present on VI. Antenna a little longer than 0.5 × body; segm. III 0.9–1.3 × segm. VI; processus terminalis about 0.8 × VIa. Abd. tergite VIII with small incisure (Fig. 167). 1.3–2.3 mm.

Distribution. In Sweden known from Sk., Öl., Gtl., Upl., Vstm., and Jmt.; not in Denmark, Norway, or Finland. – Great Britain, Germany, Poland.

Biology. The species lives on *Carex panicea, flacca,* and *hostiana,* especially in wet places in woodland (Stroyan 1972). Ossiannilsson collected oviparae in September in central Sweden.

109. ***Subsaltusaphis picta*** (Hille Ris Lambers, 1939)
Fig. 168.

Saltusaphis picta Hille Ris Lambers, 1939b: 106. – Survey: 414.

Apterous viviparous female. Whitish yellow, with rather dark margins, two rather dark pleural longitudinal rows of segmental spots, a similar short median row on mid-dorsum, and darker intersegmental transverse stripes or rows of muscle sclerites. Antennae black with segm. I–II, and basal half of III, pale. Tarsi brownish. Dorsum with numerous nodules that become finer and sparser towards margins as well as towards mid-dorsum; with numerous mushroom-shaped hairs all over dorsum and margins. Normal hairs present on margins of abd. segm. VI–VIII, more rarely only on VII–VIII. Antenna about 0.6 × body; segm. III 1.2–1.5 × segm. VI; processus terminalis 0.8–1.1 × VIa. Posterior angle of abd. tergite VII obtuse, about 120°. Abd. tergite VIII slightly emarginate (Fig. 168). 2.4–3.0 mm.

Alate viviparous female. Head, thorax, and cauda almost black. Abdomen with central shield on tergites III–V, and smaller segmental sclerites on other segments, dusky. Also dark spots around siphunculi, and intersegmental transverse rows of muscle sclerites present. Ant. segm. III with 11–16 rhinaria. Abd. tergite VIII more rounded than in apterae, slightly concave posteriorly.

Oviparous female. Proximal two thirds of hind tibia slightly swollen, rather dark, with about 100 scent plaques. Otherwise very much like apterous viviparous female.

Apterous male. Body slender. Antenna about 0.8 × body; processus terminalis about 0.6–0.7 × VIa; secondary rhinaria on ant. segm. III: 15–17, IV: 5–6, V: 6–8, VIa: 6–9. About 2.2 mm.

Distribution. In Sweden from Sk. in the south to Hrj. in the north; in Finland in Sa; not in Denmark and Norway. – Widespread in Europe, south to Austria, Czechoslovakia, and Hungary, east to S Russia; it is known from Great Britain, N Germany, and Poland (not the Baltic region); not in NW & W Russia.

Biology. The species lives on *Carex acuta,* possibly also *acutiformis* and other *Carex* spp., in wet localities. It has now and then been found on *Scirpus* spp. Ossiannilsson found sexuales in September and October in Sweden.

110. ***Subsaltusaphis rossneri*** (Börner, 1940)
Figs. 169, 181–183.

Saltusaphis rossneri Börner, 1940: 1. – Survey: 414.

Apterous viviparous female. Ochreous yellow, with darker margins, dusky pleural longitudinal stripes or segmental spots from head at least to mesonotum, and dark intersegmental transverse stripes, sometimes also with dusky median longitudinal stripe (Fig. 181). Antennae black with segm. I–II, and base of III, brownish. Legs dusky, with tarsi, and a spot on hind femur near the knee, somewhat darker. Dorsum rather evenly covered with irregular rows of rather pointed and coarse nodules (Figs. 182, 183); finer nodules present along intersegmental stripes; with rather few mushroom-shaped hairs on head and along margins, very few such hairs on mid-dorsum. Normal hairs on abd. tergites (VI–) VII and VIII. Antenna about 0.5 × body; segm. III 1.0–1.4 × segm. VI; processus terminalis 1.0–1.4 × VIa; longest antennal hair about 0.6–0.7 × IIIbd. Posterior angle of abd. tergite VII about 120°. Posterior margin of tergite VIII with small incisure (Fig. 169). 2.3–2.5 mm.

Alate viviparous female. Abdomen with dark central shield formed by fused cross bars on tergites III–V, short cross bar on tergite VI, smaller dorsal sclerites on other tergites, and marginal sclerites, those on tergite VI surrounding the siphunculi. Abd. tergite VIII very slightly emarginate at posterior margin, with shorter hairs also on disc, in addition to marginal hairs.

Distribution. In Denmark known from WJ (Grovsø at Grærup); widespread in Sweden, from Sk. north to P.Lpm.; in Norway known from HEs; in Finland from Ok. –

Great Britain, Germany (not N Germany), and Poland (including the Baltic region), and Czechoslovakia.

Biology. The species lives on *Carex rostrata, elata,* and possibly also *vesicaria.*

Genus *Saltusaphis* Theobald, 1915

Saltusaphis Theobald, 1915: 138.
Type-species: *Saltusaphis scirpus* Theobald, 1915.
Survey: 384.

Body rather elongate. Wax pores absent. Dorsal hairs short, fan-shaped. Such hairs present also between eye and antennal socket in some species (incl. *lasiocarpae*). Accessory rhinaria on ant. segm. VI isolated as in *Iziphya.* Fore and middle legs modified for leaping, with enlarged femora and sclerotic "knee-caps" (Fig. 186) as in *Iziphya.* Empodial hairs spatulate. Siphunculus (Fig. 187) conical, about as high as its apical diameter, placed on anterior part of segm. VI. Abd. tergite VIII more or less incised.

Four species in the world, one in Scandinavia.

Key to species of *Saltusaphis*

Apterous viviparous females

1 Processus terminalis longer than 1.3 × VIa. Antenna usually shorter than 0.9 × body. On *Carex hirta* and some other Cyperaceae .... *scirpus* Theobald
– Processus terminalis about 1.3 × VIa. Antenna about 0.9 × body. On *Carex lasiocarpa* .... 111. *lasiocarpae* (Ossiann.)

111. ***Saltusaphis lasiocarpae*** (Ossiannilsson, 1953)
Figs. 185–188.

Hiberaphis lasiocarpae Ossiannilsson, 1953: 234. – Survey: 385.

Apterous viviparous female. Greenish yellow, with wax powder. Antennae black from basal part of segm. III to VIa. Siphunculi on dark sclerites. Dorsal hairs short, fan-shaped, placed on brownish scleroites, which are more or less fused into longitudinal bands. Abd. tergites VI–VIII with relatively long, blunt or capitate hairs; abd. segm. VIII posteriorly with two large tubercles, with two long hairs or more. Antennae about 0.9 × body; processus terminalis about 1.3 × VIa, 0.4 × segm. III; antennae without secondary rhinaria; antennal hairs shorter than 0.5 × IIIbd. Apical segm. of the short rostrum an almost equilateral triangle. About 2.5 mm.

Oviparous female. Similar to apterous viviparous female. Hind tibiae slightly swollen, with numerous scent plaques.

Apterous male. Orange. Rather slender, almost parallel-sided. Antennae about 1.1 × body; secondary rhinaria on segm. III: 20–33, IV: 5–15, V: 8–16. In some individuals the dorsal hairs are less distinctly fan-shaped than in females, more club-like or rod-shaped. Rather small, about 1.7 mm.

Distribution. Only known from Sweden: Upl., Ång., Vb., and Nb.

Biology. The host is *Carex lasiocarpa*. Sexuales have been found from August to October i Uppland.

Note. The differences between *lasiocarpae* and *scirpus* are so small that the determination depends on knowledge of the host plant. Future studies may well show that *lasiocarpae* is a subspecies or even a synonym of *scirpus*.

Saltusaphis scirpus Theobald, 1915
Figs. 189–191.

Saltusaphis scirpus Theobald, 1915: 138.
Hiberaphis iberica Börner, 1949: 52.
Survey: 385.

Apterous viviparous female. Greyish yellow, sometimes a little greenish, with wax powder. Much like *lasiocarpae*, except that processus terminalis is relatively longer, 1.6–2.2 × VIa, 0.4–0.5 × segm. III. Antenna 0.7–0.9 × body. 2.0–2.3 mm.

Fundatrix. Thicker than apterous viviparous females of later generations; antenna shorter, about 0.6 × body; rod-shaped hairs only on abd. tergite VIII.

Alate viviparous female. Abdomen with transverse dorsal, more or less fused bands on segm. III–VIII. Hairs rod-shaped. Antenna 0.8–0.9 × body; processus terminalis 1.9–2.3 × VIa; segm. III with 10–21 secondary rhinaria. Wing veins bordered with brown. 1.8–2.2 mm.

Oviparous female. Hind tibia slightly swollen, with about 60 scent plaques. Similar to apterous viviparous female, sometimes somewhat larger.

Apterous male. Yellow, without wax powder. Antenna about 0.95 × body; secondary rhinaria on segm. III: 22–25, IV: 12–18, V: 6. About 1.6 mm.

Distribution. Not found in Scandinavia, but perhaps overlooked. C, E & S Europe, Middle East, C Asia, Africa. Introduced in N America (Ohio).

Biology. Recorded from various Cyperaceae; in Germany and Poland in particular found on *Carex hirta*.

Genus *Nevskyella* Ossiannilsson, 1954

Nevskya Ossiannilsson, 1953: 233, praeocc.
Type-species: *Nevskya fungifera* Ossiannilsson, 1953.
Nevskyella Ossiannilsson, 1954a: 54, replacement name.
Survey: 320.

Body oval, convex. Dorsum covered with minute spinules arranged in transverse rows. Dorsal protuberances or tubercles present only on posterior part of abd. segm. VIII, each with a spatulate or almost fan-shaped hair. Legs and cauda with normal hairs.

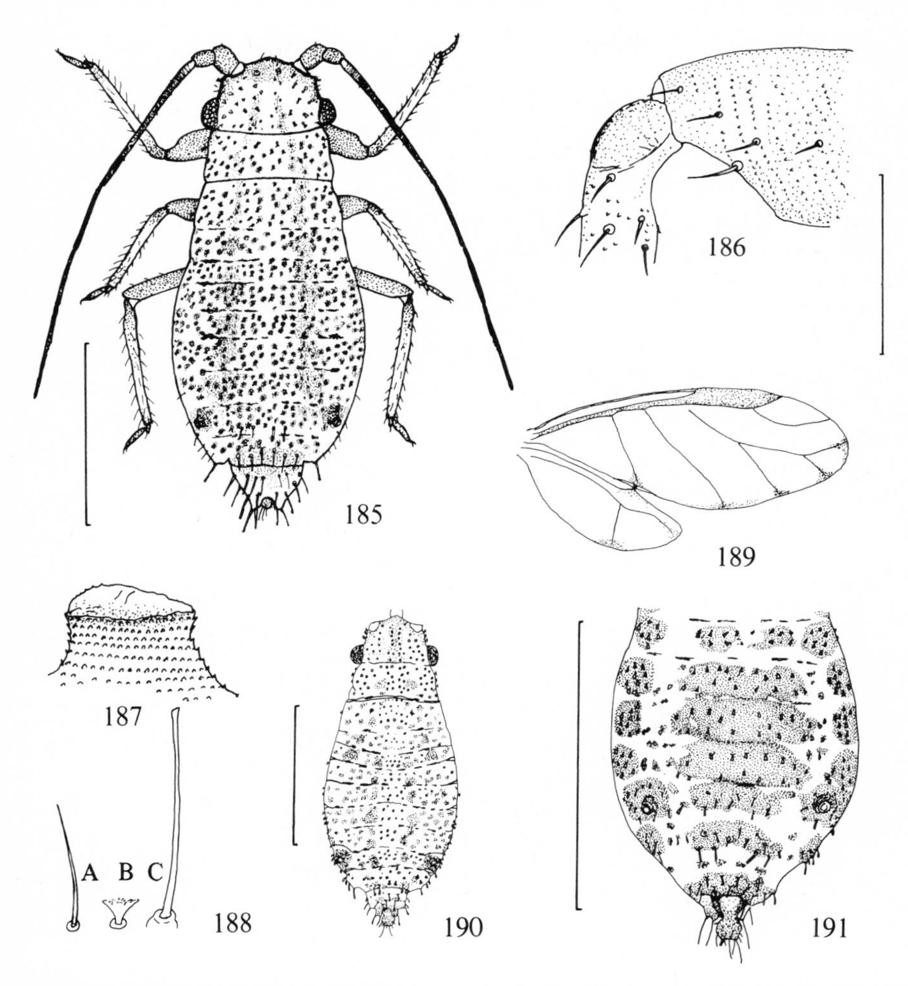

Figs. 185–188. *Saltusaphis lasiocarpae* (Ossiann.), apt. viv. (paratype). – 185: habitus; 186: knee of fore leg; 187: siphunculus; 188: types of body hairs; A: pointed (on frons and venter), B: fan-shaped (dorsum), C: blunt or slightly capitate (posterior part of abdomen). (Scale 1 mm for 185).
Figs. 189–191. *S. scirpus* Theobald. – 189: wings; 190: body of apt. viv.; 191: abdomen of al. viv. (Scales 1 mm for 190 and 191). (189–191 after Szelegiewicz, redrawn).

Dorsal hairs mushroom-shaped, rather numerous and evenly scattered. Accessory rhinaria on ant. segm. VI placed closely to primary rhinarium as in most other genera, but in contrary to the condition found in *Iziphya* and *Saltusaphis*. Shape of legs, and shape and arrangement of siphunculi, as in *Iziphya*. Empodial hairs spatulate.

Two species in the world, one species in Scandinavia.

112. ***Nevskyella fungifera*** (Ossiannilsson, 1953)
Figs. 192–196.

Nevskya fungifera Ossiannilsson, 1953: 233. – Survey: 320.

Apterous viviparous female. Yellow, sometimes with large dark areas on both sides of the dorsal mid-line, sometimes more or less uniformly brownish pigmented, and with only traces of the typical colour pattern. Dorsal hairs mostly mushroom-shaped. Antenna about 0.7 × body; processus terminalis about 1.2–1.3 × VIa; secondary rhinaria absent. About 1.7 mm.

Fundatrix. With shorter antenna, about 0.6 × body; processus terminalis about 1.0 × VIa.

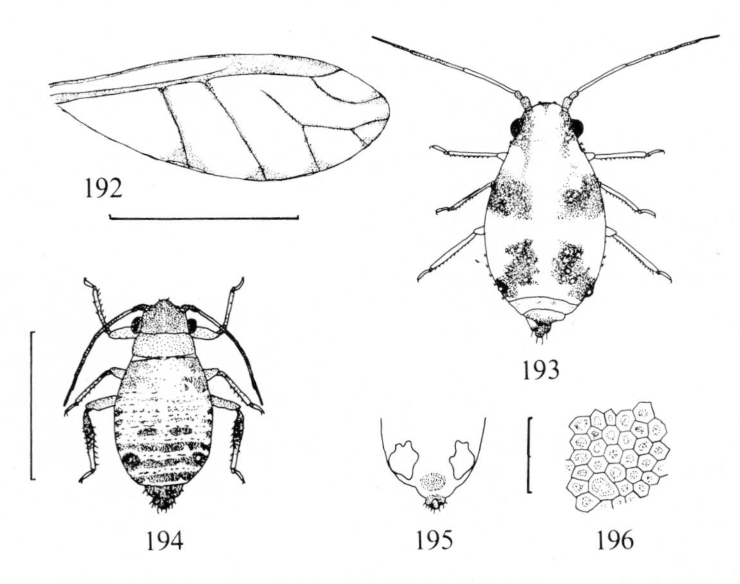

Figs. 192–196. *Nevskyella fungifera* (Ossiann.). – 192: fore wing; 193: apterous viviparous female; 194: oviparous female; 195: posterior part of abdomen of oviparous female, ventral view, with subsiphuncular wax gland plates in outline; 196: part of subsiphuncular wax gland plate of ovip., with honeycomb-pattern. (Scales 1 mm for 192, 194 & 195, 0.02 mm for 196). (193 after Ossiannilsson, redrawn; others orig.).

Alate viviparous female. Head and thorax dark; head with paler median longitudinal stripe. Abdomen with dark marginal and dorsal sclerites. Antenna about 0.9 × body; processus terminalis 0.9 × 1,3 × VIa; secondary rhinaria on segm. III: 9–17, IV: 0–2. Vein of fore wing dark-bordered (Fig 192).

Oviparous female. Colour as in apterous viviparous female. Hind tibiae swollen, with 30–40 scent plaques (Fig. 194).

Apterous male. Yellow with fuscous head and four large patches on dorsum. Slender. Secondary rhinaria on ant. segm. III: 11–17, IV: 3–7, V: 4–7.

Distribution. Only known from Sweden, Upl.: Vaksala, Jälla.

Biology. The species lives on *Carex caryophyllea* (= *verna*). Oviparae occur in September. Fundatrices appear in May.

Genus *Iziphya* Nevsky, 1929

Iziphya Nevsky, 1929b: 314.
Type-species: *Iziphya maculata* Nevsky, 1929.
Juncobia Quednau, 1954: 40.
Type-species: *Iziphya leegei* Börner, 1940.
Survey: 228.

Body oval or pear-shaped, rather broad. Apterae usually with dorsal pattern of pigmented blackish areas; dorsum without membranous borders between segments from mesonotum to abd. tegite I and from abd. tergite (III–)IV to VI. Abdomen of alatae with a more or less regular series of dark bands on tergites III–VIII, those on III–VI often enlarged and fused into a central shield, which may be united with siphuncular sclerites. Dorsal wax gland groups absent. Dorsal hairs usually mainly short and fan-shaped (flabellate) (Fig. 217); other kinds of hairs occur, but not mushroom-shaped hairs. Short, fan-shaped hairs not present on lateral margins of head between eye end antennal socket (in contrary to some species of *Saltusaphis*). Tibiae usually with various kinds of hairs: flabellate, rod-shaped, capitate, pointed. Cuticle with rows of minute spinules. Frons convex and rather broad. Accessory rhinaria on ant. segm. VI isolated, not close to primary rhinarium (Fig. 212). Apterae without secondary rhinaria. Rostrum short. Fore and middle legs modified for leaping, with enlarged femora and thick bases of tibiae; base of tibia with smooth, heavily sclerotized "knee-cap" (Fig. 213), which articulates with femur. Empodial hairs spatulate. Fore wings with dark-bordered veins and dark areas along distal parts of veins; hind wing with one oblique vein. Siphunculus stump-shaped with rounded apical rim, apparently placed between segm. V and VI (Fig. 216), with rows of denticles. Dorsal tubercles present on abd. segments in some species, usually best developed on posterior segments. Abd. tergite VIII not, or only apparently, incised, because of presence of tubercles.

Twenty species in the world, five species in Scandinavia.

Key to species of *Iziphya*

Apterous viviparous females

1 All dorsal and marginal hairs slightly capitate or pointed (Figs. 199, 203). .... 115. *ingegardae* Hille Ris Lambers

– At least some of the dorsal and marginal hairs short and fan-shaped 2

2(1) Frontal hairs and dorsal hairs short and fan-shaped. Margins of head between eye and antennal socket sloping strongly toward frons. Processus terminalis shorter than VIa. .... 116. *leegei* Börner

– Frontal hairs and some of the dorsal hairs not fan-shaped. Margins of head between eye and antennal socket almost parallel. Processus terminalis longer than VIa. 3

3(2) With a few rod-shaped marginal hairs anterior to siphunculi (Fig. 204). On *Carex stellulata* and *praecox*. .... 117. *memorialis* Börner

– Without rod-shaped marginal hairs anterior to siphunculi. On *Carex arenaria* or *canescens*. 4

4(3) On *Carex canescens*. .... 113. *austriaca* Börner

– On *Carex arenaria*. .... 114. *bufo* (Walker)

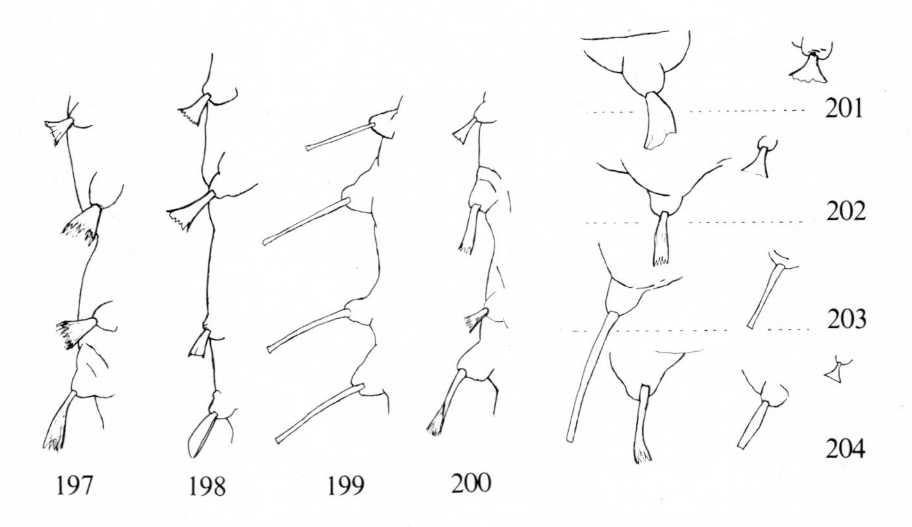

Figs. 197–204. Shape of hairs of apt. viv. of *Iziphya* spp. – 197–200: marginal hairs of abd. segm. II and III; 201–204: dorsal hairs of abd. segm. III. – 197, 201: *austriaca* Börner; 198, 202: *bufo* (Wlk.); 199, 203: *ingegardae* H. R. L.; 200, 204: *memorialis* Börner.

113. ***Iziphya austriaca*** Börner, 1950
Figs. 197, 201, 205, 206.

Iziphya austriaca Börner, 1950: 4.
Iziphya suecica Hille Ris Lambers, 1952b: 57.
Survey: 228.

Apterous viviparous female. Colour and most other characters as in *bufo*. Hairs on dorsum and margins short and fan-shaped (Figs. 197, 201); longer fan-shaped or rod-shaped hairs usually only present on abd. tergites VII–VIII; pointed hairs present on frons, antennae, legs, and cauda. Spinal hairs placed on tubercles. Dorsal cuticle without distinct spinules except on and around bases of siphunculi, on abd. tergites VII–VIII, and on appendages and cauda. Antenna 0.7–0.8 × body; processus terminalis 1.5–1.8 × VIa. 1.4–2.1 mm.

Alate viviparous female. Dorsal hairs mostly fan-shaped, but more slender than in the apterous viviparous female. Ant. segm. III with about 19–21 rhinaria, IV with 0–2. Wing veins broadly dark-bordered (Fig. 206).

Distribution. In Sweden known from Sk., Sdm., Upl., Vstm., Vrm., and Vb.; in Finland from N and Ok; not in Denmark or Norway. – Scotland, the Baltic region of Poland, Austria, and Ukraine.

Biology. The species lives in wet places on the leaves of *Carex canescens*.

Note. The description is mainly based on Börner (1950) and Szelegiewicz (1976). The species is very closely related to *bufo*. Szelegiewicz could not separate the two species on basis of morphological characters, which are constant in both species, and I cannot either. The morphological difference given by Börner regarding presence or absence of rod-shaped dorsal hairs on the tergites anterior to abd. segm. VII, does not exist in Scandinavian material.

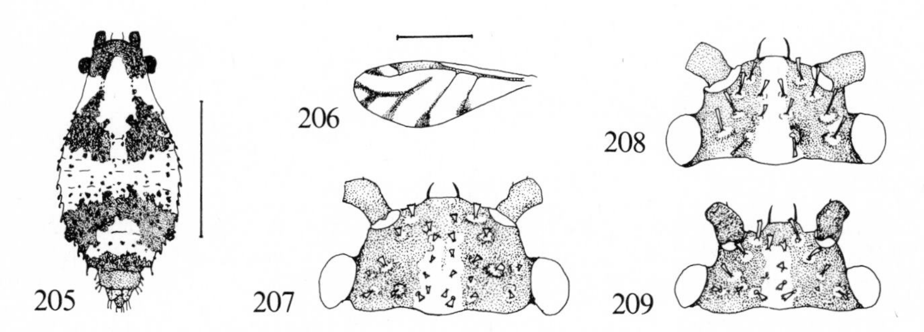

Figs. 205, 206. *Iziphya austriaca* Börner. – 205: body of apt. viv.; 206: fore wing. (Scales 1 mm). (After Szelegiewicz, redrawn).
Figs. 207–209. Heads of *Iziphya* spp. – 207: *bufo* (Wlk.); 208: *ingegardae* H. R. L.; 209: *memorialis* Börner. (After Szelegiewicz, redrawn).

114. ***Iziphya bufo*** (Walker, 1848)
Plate 1: 5. Figs. 198, 202, 207, 210–213.

Aphis bufo Walker, 1848b: 46. – Survey: 228.

Apterous viviparous female. Yellow with blackish grey dorsal markings, usually with two large dark areas on metathorax, dark areas around siphunculi more or less joined by a dark cross band, dark tergites VII–VIII, and numerous small sclerites. Head (Fig. 207) dark with paler central area. Ant. segm. I–II and IV–VI dark, segm. III pale with dark apex. Eyes, femora, bases of tibiae, tarsi, and siphunculi, black. Dorsum and margins with short fan-shaped hairs (Figs. 198, 202) and longer capitate or rod-shaped hairs on abd. tergites (V–)VI–VIII; spinal hairs placed on dark tubercles or tubercular bases; pointed hairs present on head, antennae, legs, and cauda. Antenna about 0.7–0.8 × body; processus terminalis about 1.5 × VIa; longest hair on segm. III shorter than 0.5 × IIIbd. 1.6–2.0 mm.

Alate viviparous female. Middle abd. tergites each with two irregular transverse rows of hairs. Ant. segm. III with about 18–20 rhinaria along entire segment.

Oviparous female. Similar to the apterous viviparous female, but dorsal dark pattern more broken. Hind tibiae swollen, with scent plaques.

Distribution. In Denmark known from WJ, NWJ, NEJ, and NEZ, especially common in dune areas along the west coast of Jutland; in Sweden known from Sk. in the south to Med. in the north; not in Norway or Finland. – Widespread in Europe, south at least to Hungary; widespread in Great Britain, mainly coastal; common in N Germany; known from the Baltic region of Poland.

Biology. The host plants are *Carex arenaria* and – according to Szelegiewicz (1976 and 1977) – also *C. ligerica* and *C. caryophyllea.*

115. ***Iziphya ingegardae*** Hille Ris Lambers, 1952
Figs. 199, 203, 208, 215.

Iziphya ingegärdae Hille Ris Lambers, 1952b: 57. – Survey: 228.

Apterous viviparous female. Much like *bufo,* but all hairs on dorsum and margins long, slender, and slightly capitate or pointed (Figs. 199, 203). Antenna about 0.8 × body; processus terminalis about 1.7–2.0 × VIa. About 2.1–2.2 mm.

Distribution. In Sweden known from Sdm. (with the type locality), Upl., Vstm., and Vrm.; not in Denmark, Norway, or Finland. – Germany (Berlin), Poland (rare; not in the Baltic region).

Biology. The species lives on *Carex canescens* and *leporina.*

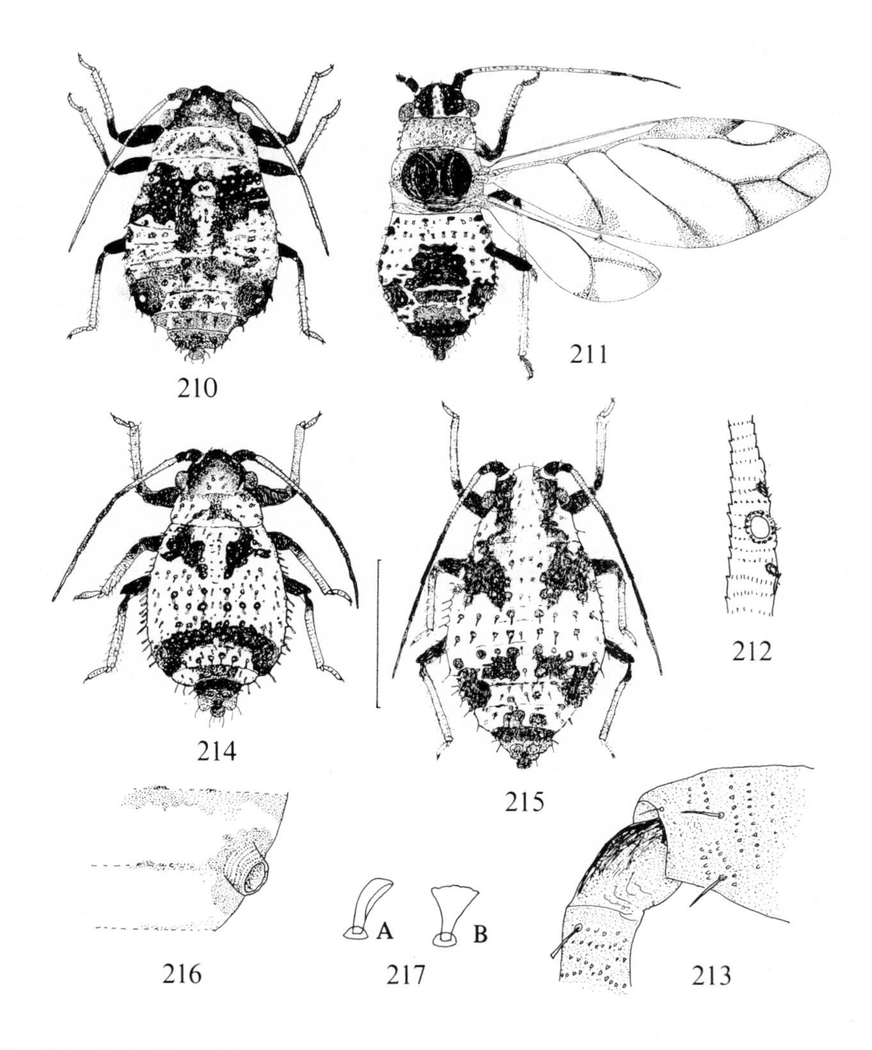

Figs. 210–213. *Iziphya bufo* (Wlk.). – 210: apt. viv.; 211: al. viv.; 212: part of ant. segm. VI of al. viv.; 213: knee of fore leg of apt. viv.
Fig. 214. *I. memorialis* Börner, apt. viv.
Fig. 215. *I. ingegardae* H. R. L., apt. viv.
Figs. 216, 217. *I. leegei* Börner, apt. viv. – 216: right margin of abd. segm. V and VI, with siphunculus (hairs not drawn); 217: marginal hairs of abd. segm. VI, seen from the edge (A), and from the flat side (B). (Scale 1 mm for 211, 214, 215). (211 after Szelegiewicz, redrawn; the others orig.).

116. ***Iziphya leegei*** Börner, 1940
Plate 1: 4. Figs. 216, 217.

Iziphya leegei Börner, 1940: 1. – Survey: 228.

Apterous viviparous female. Yellowish with grey and black dorsal pattern of markings similar to that of *bufo*. Colour of antennae, legs, and siphunculi as in *bufo*. Eyes red. All hairs on dorsum and margins short and fan-shaped, including the hairs on head; dorsal hairs not on prominent tubercles. Antennal and caudal hairs, and some hairs on the legs, pointed. Anterior edge of head almost semicircular. Antenna about 0.5 × body; processus terminalis about 0.7–0.8 × VIa; longest hair on segm. III about 0.5 × IIIbd. 1.6–1.9 mm.

Alate viviparous female. Middle abd. tergites, at least partly, with three irregular transverse rows of hairs. Ant. segm. III with about 10 rhinaria.

Oviparous female. Similar to the apterous viviparous female, but dorsal dark pattern more broken. Hind tibiae swollen, with scent plaques.

Apterous male. Rather similar to viviparous females. Secondary rhinaria on ant. segm. III: 10–16, IV: 5–8, V: 5–7. Rather small, about 1.2–1.3 mm.

Distribution. In Denmark found in NWJ and NEZ; in Sweden from Sk. in the south to Hls. in the north; in Norway known from HOy and HOi; not in Finland. – Great Britain, the Netherlands, Germany (not N Germany), France, Spain, Czechoslovakia, Hungary, Yugoslavia, and S Russia.

Biology. The species lives on *Juncus gerardi, compressus,* and *maritimus.*

117. ***Iziphya memorialis*** Börner, 1950.
Figs. 200, 204, 209, 214.

Iziphya memorialis Börner, 1950: 4. – Survey: 228.

Apterous viviparous female. Similar to *bufo,* but rod-shaped hairs present on abd. tergites (III–)V–VIII (Fig. 204), also on margins anterior to siphunculi. 1.6–1.9 mm.

Distribution. In Sweden found in Upl. and Vstm.; not in Denmark, Norway, or Finland. – Germany (but not in N Germany), Austria, Czechoslovakia, and Poland (but not in the Baltic region).

Biology. The species lives on *Carex stellulata* and *caryophyllea* (=*praecox*).

SUBFAMILY CHAITOPHORINAE

Viviparous females apterous or alate; fundatrices apterous. Body usually with numerous long hairs, which may be pointed, blunt, spatulate, or furcate. Dorsum without wax glands. Frons convex or straight, without tubercles. Antennae 4-, 5-, or 6-segmented; primary rhinaria of alatae sometimes surrounded by short hairs; accessory rhinaria on ultimate ant. segm. placed close to the primary rhinarium, sometimes more distinctly fringed; secondary rhinaria circular, usually missing in apterous females. Rostrum usually rather short, not telescopic; segment II without sclerotized wishbone-shaped arch. Tarsi without spinules, but tibiae sometimes with fine spinules between the hairs over part of their length. Empodial hairs simple or rod-shaped, sometimes with flattened apices. Siphunculi short, truncate, often reticulate (tribe Chaitophorini (Fig. 4)), sometimes pore-shaped. Cauda semicircular, or more or less distinctly knobbed. Anal plate rounded or slightly emarginate, never bilobed. With 4 rudimentary gonapophyses (Fig. 9).

Host alternation does not occur. The aphids live on deciduous trees or on herbaceous monocotyledones. Galls are never produced. Some species are visited by ants. The subfamily is subdivided into two tribes.

Key to tribes of Chaitophorinae

Apterous and alate viviparous females

1 Siphunculi with at least some trace of reticulate sculpture, more or less stump-shaped. Antennae 6-segmented. .... *Chaitophorini* (p. 107)
– Siphunculi without reticulate sculpture, stump-shaped, conical, or pore-shaped. Antennae 4– or 5–segmented. .... *Siphini* (p. 141)

TRIBE CHAITOPHORINI

Body oval or somewhat elongate, long-haired. Antennae shorter than body, 6–segmented. Siphunculi low, truncate, subcylindrical or stump-shaped, with reticulate apical part.

The hosts are deciduous trees, especially Salicaceae and Aceraceae. Two genera occur in Scandinavia.

Key to genera of Chaitophorini

Apterous and alate viviparous females

1 Cauda usually distinctly knobbed. Empodial hairs simple or rod-shaped. On *Salix* or *Populus*. .... *Chaitophorus* Koch (p. 121)
– Cauda broadly rounded to slightly knobbed. Empodial hairs somewhat widened towards apices. On *Acer* or *Aesculus*. .... *Periphyllus* van der Hoeven (p. 107)

Genus *Periphyllus* van der Hoeven, 1863

Periphyllus van der Hoeven, 1863: 7.

Type-species: *Periphyllus testudo* van der Hoeven, 1863 = *Phyllophora testudinacea* Fernie, 1852.
Survey: 348.

Body elongate oval or pear-shaped. Dorsum of apterae mainly membranous with hair-bearing small sclerites. Abdomen of alatae with dark segmental dorsal cross bars, or sclerites, and more or less developed marginal sclerites, sometimes also with small intersegmental muscle sclerites. Processus terminalis relatively long, from about as long as VIa in fundatrices up to about 7 × VIa in alate males; antennal hairs long, erect, usually pointed, with very thin apices; VIa usually with two hairs; alatae and sometimes also apterae (alatiform individuals occurring in the spring) with secondary rhinaria on ant. segm. III; alate males also with many secondary rhinaria on segm. IV and V. First tarsal segm. usually with 5, 6 or 7 hairs, with individual variation from 2 to 9. Empodial hairs slightly spatulate. Siphunculus slightly longer than wide at middle, stump-shaped, with well developed flange or apical flare, reticulation more extensive in alatae than in apterae (Figs. 4, 231, 233). Cauda broadly rounded or slightly knobbed; in the latter case with basal part not widening very much towards base, usually with about 12 hairs. Scent plaques on hind tibiae of oviparous females circular with small central pore.

With 29 species in the world, 6 species in Scandinavia; 3 additional species occurring in C Europe are included in the keys and descriptions below; they may have been overlooked in our area. All species feed on *Acer,* rarely also *Aesculus.* In summer, when the sieve tube sap of the trees is deficient in amino acids, some species survive exclusively as small dimorphs, or "summer larvae". These are first instar nymphs, and are very different from first instar nymphs of other generations of the same species, and also very different from the following instars of the same individual. Their intake of sap is insignificant, or does not take place at all, so they do not grow and remain first instar nymphs for several months, from May or June until September, or even later.The dimorphs of some species (Figs. 234–238) are flat and armoured with dorsal sclerotic plates much like tortoises. Leaf-shaped hairs are present around margins of head and body, and on parts of antennae and legs. The body seems bordered with these hairs, because the appendages are placed close to the body. Apparently these individuals do not move for months. The dimorphs of some other species (Fig. 223–225) are densely covered with very long hairs, and they spend the summer in close aggregations on the undersides of leaves (Fig. 224), while the dimorphs of the first mentioned species spend the summer scattered on the upper- and undersides of leaves (Fig. 235). The various adapted characters of the dimorphs can be interpreted as protection against evaporation.

The dimorphs grow up in autumn and finally become apterous viviparous sexuparae. A few species of *Periphyllus* do not produce dimorphs, while others produce both dimorphs and normal first instar nymphs during summer.

Sexuales are produced in autumn. The males are alate. The eggs hibernate on *Acer* and hatch in spring. The fundatrices are thick, with relatively short antennae and short processus terminalis. Their offspring consists partly of apterae, partly of alatae. Large colonies are formed in spring, usually visited by ants. Colonies of adult inviduals of the species, which are aestivating as dimorphs, usually disappear in early summer. The dimorphs are born by individuals of 3rd generation, or by the last-born individuals of 2nd generation, perhaps also by individuals of later generations. They can be present on the leaves in huge numbers, but are easily overlooked and never visited by ants.Colonies of these species reappear in autumn, when the dimorphs have grown up and begun their reproductive activity.

Note. Keys and descriptions are partly based on Börner (1952), Hille Ris Lambers (1966), and Stroyan (1977).

Key to species of *Periphyllus*

Apterous viviparous females (excl. fundatrices)

1 Longer hair on VIa more than 4 times as long as shorter hair on VIa (Figs. 219 A & B). Processus terminalis longer than 3 × VIa. 2

– Longer hair on VIa less than 4 times as long as shorter (or shortest) hair on VIa, or if more than 4 times as long, then processus terminalis shorter than 3 × VIa. 3

2(1) Dorsal hairs sharply pointed. Shorter hair on VIa at least as long as IIIbd. Yellowish with dark lyre-shaped mark on dorsum .... 122. *lyropictus* (Kessler)

– Dorsal hairs rather fine, but more or less blunt. Shorter hair on VIa about 0.5 × IIIbd. Green. .... 121. *hirticornis* (Walker)

3(1) Apical segm. of rostrum with more than 15 hairs (Fig. 221B). ..... *singeri* (Börner)

– Apical segm. of rostrum with less than 15 hairs. 4

4(3) Processus terminalis 3.2–4.8 × VIa (Figs. 218 C & E). Longer hair on VIa longer than VIa, more than 8 times as long as diameter of processus terminalis. 5

– Processus terminalis 3.7 × VIa or shorter. Longer hair on VIa as long as or shorter than VIa. If processus terminalis is longer than 3.1 × VIa (Fig. 218D), then longer hair on VIa less than 3 times as long as diameter of processus terminalis, shorter than VIa (Fig. 219D) 6

5(4) Siphunculus 0.3–0.5 × 2sht., dark with paler base. On *Acer platanoides*. .... 120. *coracinus* (Koch)

– Siphunculi as long as 2sht., mostly rather evenly pigmented. On *Acer campestre*. .... *obscurus* Mamontova

6(4) Knees black. 7

– Knees pale or very nearly so (Fig. 222 D). 8

7(6) Hind tibia dark only at base and apex (Fig. 222 B); hind tarsus dark. Longer hair on VIa less than 3 times as long as shorter hair on VIa. .... 123. *testudinaceus* (Fernie)

– Hind tibia dark all over; hind tarsus paler than tibia. Longer hair on VIa more than 3 times as long as shorter hair on VIa. .... *californiensis* (Shinji)

8(6) Cauda with at most 20 hairs. Longer hair on VIa 2.25–3.5 times as long as shorter hair on VIa, 7–13 × diameter of processus terminalis (Fig. 219 G). On *Acer pseudoplatanus*. ..... 118. *acericola* (Walker)
– Cauda with more than 20 hairs. Longer hair on VIa 1.4–3.2 times as long as shorter hair on VIa, 3–8 × diameter of processus terminalis (Fig. 219 H). On *Acer platanoides*. .... 119. *aceris* (Linné)

Alate viviparous females
(except *P. coracinus* and *P. obscurus*)

1 Apical segm. of rostrum with more than 4 accessory hairs. .... *singeri* (Börner)
– Apical segm. of rostrum with 0–4 accessory hairs. 2
2(1) Longer hair on VIa more than 4 times as long as shorter hair on VIa. Processus terminalis longer than 3 × VIa. 3

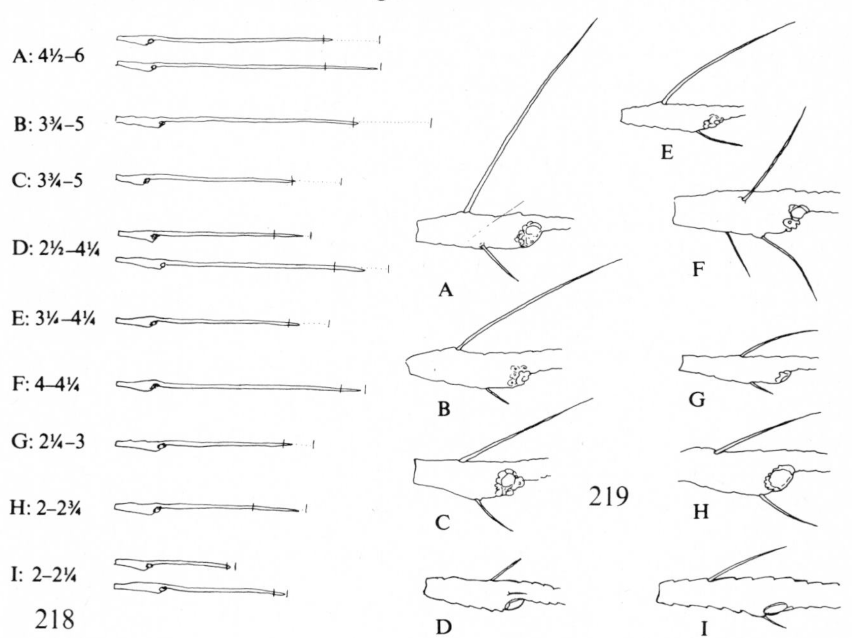

Figs. 218, 219. Ant. segm. VI of *Periphyllus* spp. – 218: the ratio VIb/VIa; the numbers and the short vertical lines connected by dots show the variation; hairs are omitted. – 219: basal part of ant. segm. VI showing proportional length of hairs; A: *lyropictus* (Kessl.); B: *hirticornis* (Wlk.); C: *coracinus* (Koch); D: *testudinaceus* (Fern.) (218: 2½–3¾ in apt. viv.); E: *obscurus* Mamont.; F: *singeri* (Börn.); G: *acericola* (Wlk.); H: *aceris* (L.); I: *californiensis* (Shinji). (A–D and F–H after Szelegiewicz, E after Stroyan, redrawn).

– Longer hair on VIa less than 4 times as long as shorter hair on VIa, or if more than 4 times as long, then processus terminalis shorter than 3 × VIa. 4

3(2) Apex of hind tibia distinctly darker than middle part. Shorter hair on VIa 0.019–0.025 mm. Siphunculus 0.21–0.28 mm. 121. *hirticornis* (Walker)

– Apex of hind tibia not distinctly darker than middle part. Shorter hair on VIa 0.025–0.040 mm. Siphunculus 0.17–0.23 mm .... 122. *lyropictus* (Kessler)

4(2) Pterostigma not black, paler than abdominal cross bars. Hind tibia dark at least apically. 5

– Pterostigma black, about equally dark with abdominal cross bars. Hind tibia pale. 6

5(4) Longer hair on VIa less than 0.05 mm long. Cross bar on abd. tergite V with 6 hairs. .... 123. *testudinaceus* (Fernie)

– Longer hair on VIa more than 0.05 mm long. Cross bar on abd. tergite V with about 10 hairs or more. .... *californiensis* (Shinji)

6(4) Cauda with at most 20 hairs. Shorter hair on VIa reaching the primary rhinarium of ant. segm. VI or shorter. Longer hair on VIa 2.5–3 times as long as shorter hair. Marginal sclerites often distinctly paler than the rather broad abdominal cross bars. .... 118. *acericola* (Walker)

– Cauda with more than 20 hairs. Shorter hair on VIa reaching past the primary rhinarium of ant. segm. VI. Longer hair on VIa 1.4–2.4 times as long as shorter hair. Marginal sclerites as dark as the rather narrow abdominal cross bars. .... 119. *aceris* (Linné)

Dimorphs

(unknown in *P. coracinus, lyropictus, obscurus,* and *singeri*)

1 Greenish with leaf-shaped marginal hairs. Occurring scattered on leaves 2

– Yellowish white with long simple dorsal and marginal hairs. Occurring in groups on leaves. 4

2(1) Processus terminalis longer than 2 × basal part of ultimate ant. segm. (IVa) (Fig. 238). Eyes bright red. .... 121. *hirticornis* (Walker)

– Processus terminalis not longer than basal part of ultimate ant. segm. (IVa). Eyes blackish red. 3

3(2) Abdominal dorsum with spinal, pleural, and marginal plates (Fig. 234). .... 123. *testudinaceus* (Fernie)

– Abdominal dorsum with spinal and pleuromarginal plates (Fig. 237). .... *californiensis* (Shinji)

4(1) Lateroapical hairs on second tarsal segm. hardly spatulate at apex (Fig. 225 A). Empodial hairs almost linear. On *Acer pseu-*

doplatanus. .... *118. acericola* (Walker)

– Lateroapical hairs on second tarsal segm. distinctly spatulate at apex (Fig. 225 B). Empodial hairs distinctly widened from base to apex. On *Acer platanoides.* .... 119. *aceris* (Linné)

118. ***Periphyllus acericola*** (Walker, 1848)
Figs. 218 G, 219 G, 222 C & D, 223, 224, 225 A, 226.

Aphis acericola Walker, 1848a: 451. – Survey: 348.

Apterous viviparous female. Pale green or yellowish, sometimes with dorsal brownish markings. Ant. segments with rather dark apices. Head, pronotum, legs (except tarsi), and siphunculi pale. Dorsal hairs not placed on dark sclerites. Processus terminalis 2.3–3.0 × VIa (Fig. 218 G); longer hair on VIa as long as or shorter than VIa, 7–13 times as long as diameter of processus terminalis, 2.25–3.5 times as long as shorter hair on VIa (Fig. 219 G); segm. III with 0–6 rhinaria and 12–20 hairs. Apical segm. of rostrum about as long as VIa, about 0.8 × 2sht. First tarsal segm. with 7 hairs. Siphunculus about as long as 2sht., with 3–4 transverse rows of meshes. Cauda at base more than twice as broad as long, at most with 20 hairs. About 3.3 mm.

Fundatrix. Thicker. Dorsal hairs placed on dark sclerites. Processus terminalis about as long as VIa. Ant. segm. III with about 8 hairs. About 3.9 mm.

Alate viviparous female. Marginal sclerites often rather pale and indistinct. Dorsal cross bars rather broad, dark, and distinct. Hind tibiae quite pale, without dark apices. Siphunculi dark. Pterostigma remarkably dark. Processus terminalis 2.4–2.9 × VIa; longer hair on VIa 2.5–3 times as long as shorter hair on VIa; the latter 0.02–0.04 mm, not reaching beyond primary rhinarium on segm. VI; ant. segm. III with 6–16 rhinaria and 15–29 hairs. Cross bar on abd. tergite V with 7–15 hairs. Body length 3.5 mm or less.

Dimorph. Yellowish white. Dorsum and margins with pointed hairs about 0.5 × body length (Fig. 223). Antenna about 0.5 × body, 4-segmented. Lateroapical hairs on second tarsal segm. hardly spatulate at apex (Fig. 225 A). Empodial hairs almost linear. About 0.7 mm.

Oviparous female. Dirty brownish. Hind tibiae swollen, with scent plaques.

Distribution. Common and widespread in Denmark; in Sweden known from Sk.; recorded from Norway; not in Finland. – Widespread in Europe, south at least to the Alps and Hungary; fairly common and widespread in Great Britain; common in N Germany; recorded from the Baltic region of Poland; in Russia only recorded from the southern region.

Biology. The aphids live on the undersides of leaves, leaf stems, and young shoots of *Acer pseudoplatanus.* The dimorphs are the only individuals found in summer. They are densely aggregated in groups, resembling whitish spots, on the undersides of the leaves (Fig. 224), and do not move unless disturbed.

119. ***Periphyllus aceris*** (Linné, 1761)
Figs. 218 H, 219 H, 225 B.

Aphis aceris Linné, 1761: 262.
Chaitophorus xanthomelas Koch, 1854: 1.
Survey: 348.

Apterous viviparous female. Yellow, often with green dorsal markings. Head, pronotum, legs (except tarsi), and siphunculi, pale. Dorsal hairs nor placed on dark sclerites. Processus terminalis 2.2–2.7 × VIa (Fig. 218 H); longer hair on VIa shorter than VIa, 3–8 times as long as diameter of processus terminalis, 1.4–3.2 times as long as shorter hair on VIa (Fig. 219 H). Apical segm. of rostrum about as long as VIa, about 0.8 × 2sht. First tarsal segm. with 7 hairs. Siphunculi with 4–7 transverse rows of meshes. Cauda as in *acericola,* but with more than 20 hairs. From about 1.5 mm to a little more than 3 mm.

Alate viviparous female. Abdomen with black, rather narrow, dorsal cross bars and equally dark marginal sclerites. Hind tibiae pale all over. Siphunculi dark. Pterostigma

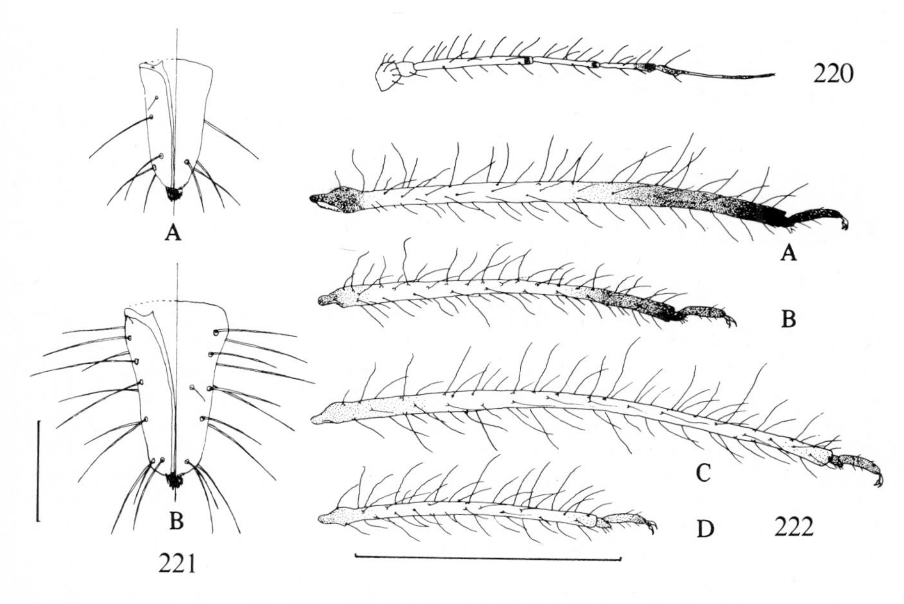

Fig. 220. Antenna of apt. viv. of *P. coracinus* (Koch). (After Szelegiewicz, redrawn).
Fig. 221. Apical segm. of rostrum of apt. viv. of *Periphyllus coracinus* Koch) (A) and *P. singeri* (Börner) (B), left half in frontal view, right half in caudal view. (Scale 0.1 mm).
Fig. 222. Hind tibia and tarsus of *P. testudinaceus* (Fern.) (A and B) and *P. acericola* (Wlk.) (C and D); A and C: al. viv.; B and D: apt. viv. (Scale 1 mm).

remarkably dark. Processus terminalis 1.9–2.5 × VIa; longer hair on VIa 1.4–2.4 times as long as shorter hair on VIa; the latter 0.025–0.1 mm, its apex reaching beyond primary rhinarium on ant. segm. VI; ant. segm. III with 5–12 rhinaria and 16–21 hairs. Cross bar on abd. tergite V with about 10–11 hairs. Body length more than 3.5 mm.

Dimorph. Similar to *acericola,* but lateroapical hairs on second tarsal segm., and empodial hairs, distinctly spatulate (Fig. 225 B).

Oviparous female. Very much like the apterous viviparous female. Ant. segm. III with 13–14 hairs. Hind tibia with basal $^{2}/_{3}$ somewhat swollen, with about 50–60 scent plaques.

Alate male. More slender than alate female. Abdominal sclerites and cross bars, and pterostigma somewhat paler than in alate female. Processus terminalis 3.1–4.0 × VIa; secondary rhinaria on ant. segm. III: 75–100, IV: about 50–80, V: about 23–35.

Distribution. In Denmark known from NWJ; in Sweden common and widespread from Sk. in the south to Dlr. and Hls. in the north; not in Norway; widespread in Finland north to ObN. – Europe, south to Spain, east to Russia; apparently not common in Great Britain; known from Germany, but not from N Germany; recorded from the Baltic region of Poland and NW & W Russia. Recorded from several states of the USA and from British Columbia in Canada.

Biology. The biology is similar to that of *acericola,* the host being *Acer platanoides.* Sexuales occur in Denmark in October–November.

Periphyllus californiensis (Shinji, 1917)
Figs. 218 I, 219 I, 227, 237.

Thomasia californiensis Shinji, 1917: 61. – Survey: 348.

Apterous viviparous female. Dark olive green to brown. Head, pronotum, legs, and siphunculi dark; hind tarsi paler than femora and tibiae. With dark marginal sclerites and dorsal sclerites or cross bars. Middle abd. tergites with a total of 10–22 dorsal hairs on each segm.; marginal sclerites with 7–12 hairs. Processus terminalis about 2 × VIa (Fig. 218 I); longer hair on VIa more than 0.05 mm long, but shorter than VIa, about 4 × diameter of processus terminalis, more than 3 times as long as shorter hair on VIa (Fig. 219 I); segm. III with about 12 hairs. Apical segm. of rostrum about as long as VIa, about 0.8 × 2sht. First tarsal segm. with 5 hairs. Siphunculi shorter than in *testudinaceus.* 2.3–3.5 mm.

Alate viviparous female. Abdomen with dark marginal sclerites and dorsal cross bars darker than pterostigma. Ant. segm. III with 7–25 rhinaria. Cross bar on abd. segm. V with more than 6 hairs.

Dimorph. Similar to *testudinaceus,* but without intersegmental pleural plates (Fig. 237).

Distribution. Not yet found in Scandinavia. – Europe, Asia, N America. In Europe probably introduced with nursery stock (Stroyan 1977).

Biology. The aphids occur on the undersides of leaves of maples which are not native in Scandinavia *(Acer palmatum, japonicum* a.o.). It is also recorded from *Aesculus californica.* Aestivating dimorphs as in *testudinaceus.*

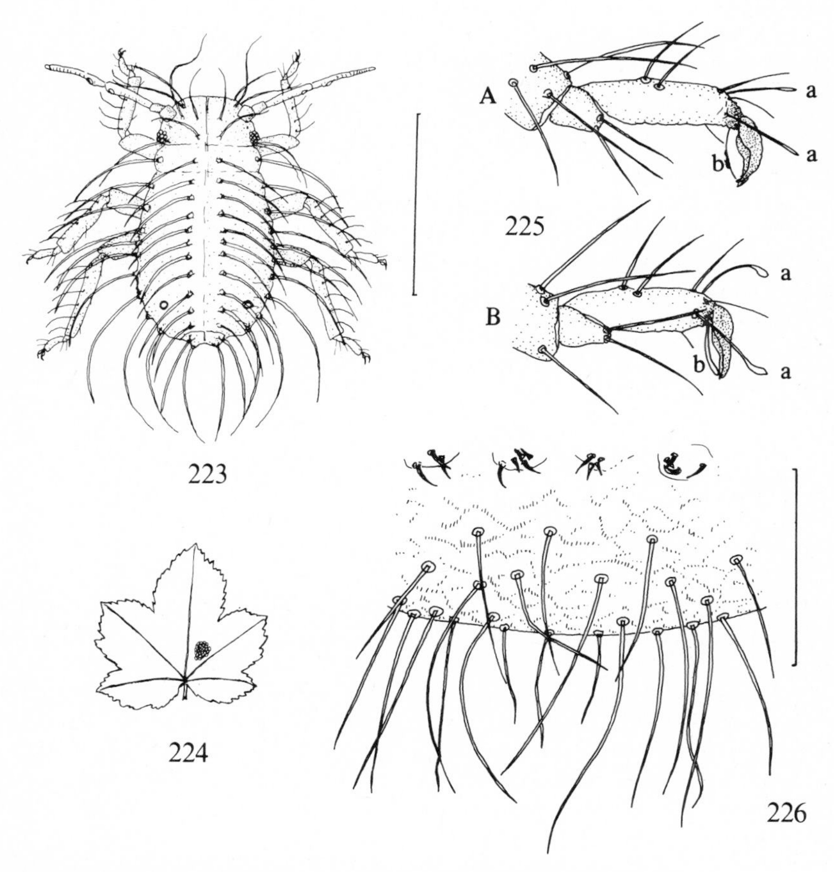

Fig. 223. Dimorph ("summer larva") of *Periphyllus acericola* (Wlk.). (Scale 0.5 mm).
Fig. 224. Dimorphs of *P. acericola* forming a cluster of twenty or more individuals on a leaf of *Acer pseudoplatanus* (see also Fig. 235).
Fig. 225. Hind tarsi of dimorphs of *acericola* (Wlk.) (A) and *aceris* (L.) (B); a = lateroapical hairs on second tarsal segm., b = empodial hairs. (Scale 0.1 mm).
Fig. 226. Posterior end of abdomen of apt. viv. of *P. acericola,* ventral view, showing 4 rudimentary gonapophyses . (Scale 0.1 mm).

120. ***Periphyllus coracinus*** (Koch, 1854)
Figs. 218 C, 219 C, 220, 221.

Chaitophorus coracinus Koch, 1854: 2. – Survey: 348.

Apterous viviparous female. Yellow, light green, dark green, brown, or mottled with green and brown. Apices of ant. segments, VIa, and siphunculi, at least distally, dark. Dorsal hairs placed on rather dark sclerites. Processus terminalis 3.7–4.8 × VIa (Fig. 218 C); longer hair on VIa longer than VIa, about 10 times as long as diameter of processus terminalis (Fig. 219 C), about 1.5 × length of shorter hair; ant. segm. III without rhinaria, with about 19–20 hairs. Apical segm. of rostrum 1.5–1.7 × VIa, about 0.9–1.0 × 2sht. First tarsal segm. with 7 hairs. Siphunculi about 0.3–0.5 × 2sht., with 3–4 transverse rows of meshes.

Alate male. Processus terminalis about 5 × VIa. Secondary rhinaria on ant. segm. III: about 80, IV: 43–48, V: about 21.

Distribution. In Sweden known from Sm. and Upl.; not in Denmark, Norway, or Finland. – W, C & E Europe, e.g. Germany, Poland, Hungary, and Russia; not recorded from Great Britain and N Germany.

Biology. The aphids live on leaf stems and twigs of *Acer platanoides*. Dimorphs are apparently not produced.

121. ***Periphyllus hirticornis*** (Walker, 1848)
Figs. 218 B, 219 B, 228, 236, 238.

Aphis hirticornis Walker, 1848b: 447.
Chaitophorus granulatus Koch, 1854: 13.
Survey: 348.

Apterous viviparous female. Light green, without dark markings (Fig. 228). Antennal and dorsal hairs, especially those on mid-dorsum, more or less blunt at apex, sometimes a little expanded, truncate, or slightly furcate. Processus terminalis about 5 × VIa (Fig. 218B); longer hair on VIa about as long as, or slightly longer than, VIa, about 8 times as long as the shorter hair; the latter only about 0.5 × IIIbd. (Fig. 219 B). Apical segm. of rostrum slightly shorter than VIa, as well as 2sht. First tarsal segm. with 6 hairs. Siphunculus longer than 2sht., strongly flared at apex, with more than 5 rows of meshes. Cauda slightly knobbed, but not broad at base. About 2–3 mm.

Alate viviparous female. Abdomen with marginal, more or less fused, spinal sclerites. Ant. segm. I–II and VI, apices of ant. segm. III–V, apices of tibiae, and tarsi, dark; knees pale. Siphunculi brownish. Cauda a little brownish. Processus terminalis about 5 × VIa; VIa with one long hair (about 0.18 mm), one shorter hair, and one very short hair close to the primary rhinarium; ant. segm. III with 2–12 rhinaria on basal $^3/_4$. Cauda as in the apterous viviparous female. About 2.4 mm.

Dimorph. Light green with leaf-like marginal hairs. Eyes bright red. Dorsum with only a median suture and intersegmental sutures (Fig. 236). Processus terminalis about 2.5 × basal part of ultimate ant. segm. (IVa) (Fig. 238).

Distribution. In Denmark known from LFM (Nakskov) and NEZ (Tåstrup); not in Sweden, Norway, or Finland. – Great Britain, the Netherlands, Germany (not N Germany), Poland, Czechoslovakia, Hungary, Bulgaria, Spain.

Biology. The aphids live on leaves, leaf stems, and fruits of *Acer campestre*. The colonies are sometimes visited by ants. Aestivating dimorphs are produced.

122. ***Periphyllus lyropictus*** (Kessler, 1886)
Figs. 218 A, 219 A, 229.

Chaitophorus lyropictus Kessler, 1886: 171. – Survey: 349.

Apterous viviparous female. Live individuals yellow with brown dorsal markings, which consist of a longitudinal stripe from head to anterior part of abdomen, and a V-shaped mark on middle of abdomen; sometimes brown all over. Apices of ant. segments, VIa,

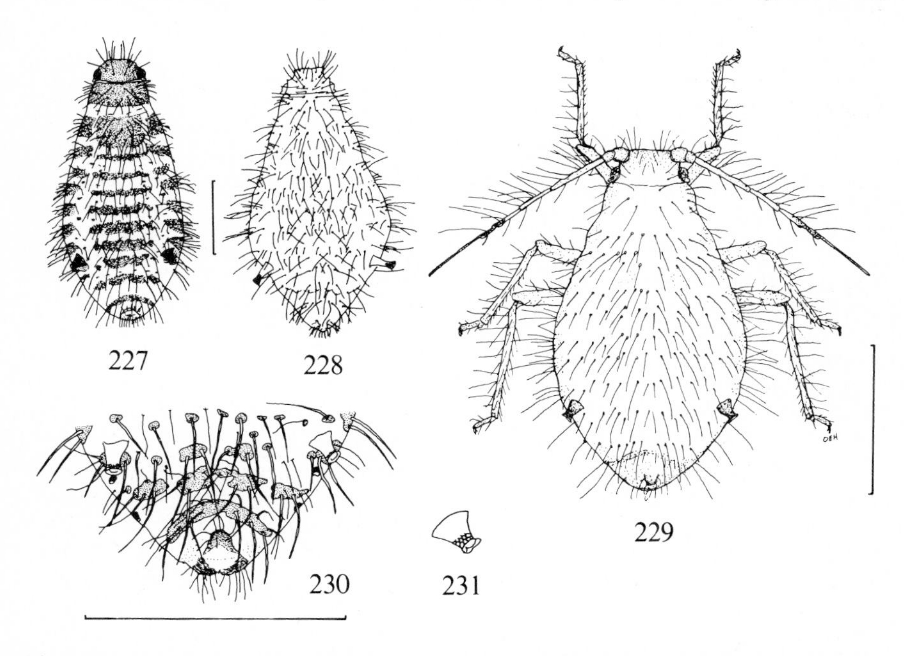

Figs. 227–229. Apt. viv. of *Periphyllus* spp. – 227: *californiensis* (Shinji); 228: *hirticornis* (Wlk.); 229: *lyropictus* (Kessl.) (from August). (Scales 1 mm). (227 and 228 after Stroyan, redrawn, 229 orig.).
Figs. 230, 231. *Periphyllus obscurus* Mamont., apt. viv. – 230: posterior end of abdomen; 231: siphunculus. (Scale 1 mm for 230). (230 after Szelegiewicz, 231 after Stroyan, redrawn).

and siphunculi, dark. Dorsal hairs very long, sharply pointed, not placed on dark scleroites. Processus terminalis about 4.5–6 × VIa (Fig. 218 A); longer hair on VIa longer than VIa, 20–25 times as long as diameter of processus terminalis, more than 4 times as long as shorter hair on VIa (Fig. 219 A); ant. segm. III with about 14–19 hairs. Apical segm. of rostrum a little longer than VIa, about 0.9 × 2sht. First tarsal segm. with 7 hairs. Siphunculi as long as 2sht., or longer, strongly flared at apex, with 3–5 transverse rows of meshes. Cauda at base narrower than 1,5 times its length 1.9–3.0 mm.

Fundatrix. Thick and large, about 3.5 mm. Processus terminalis about 2 × VIa.

Alate viviparous female. Abdomen with dark marginal sclerites and weakly developed dorsal cross bars, partly broken into smaller sclerites and not stretching out beyond the spinal region. Hind tibia of almost uniform colour, only slightly darker towards apex. Siphunculi dark. Pterostigma light brown. Processus terminalis about 5–6 × VIa; longer hair on VIa about 0.15 mm, shorter hair 0.025–0.040 mm; ant. segm. III with 6–20 rhinaria and about 18–20 hairs, the longest one about 0.22–0.25 mm.

Oviparous female. Mottled green or dark brown. Posterior part of abdomen prolonged. Processus terminalis about 5 × VIa. Hind tibia swollen, with many scent plaques.

Alate male. Similar to alate female, but more slender. Processus terminalis about 7 × VIa. Ant. segm. III, IV, and V with numerous secondary rhinaria.

Distribution. In Denmark known from SJ, EJ and NWJ; widespread all over Sweden, north to Nb.; not in Norway; in Finland known from Sa. – Great Britain (locally common, but few records), Germany (in N Germany apparently rare), Poland, NW & W Russia, and the Caucasus region; in Europe south at least to Switzerland and Hungary. N America: widespread in the USA and Canada.

Biology. The aphids live on the undersides of leaves of *Acer platanoides*. Aestivating dimorphs are not produced. The colonies are visited by ants and bees throughout the summer.

Periphyllus obscurus Mamontova, 1955
Figs. 218 E, 219E, 230, 231.

Periphyllus obscurus Mamontova, 1955. – Survey: 349.

Apterous viviparous female. Dark blackish green. Siphunculi dark. Body densely covered with long hairs. Processus terminalis 3.2–4.1 × VIa (Fig. 218 E); longer hair on VIa 0.16 mm or longer, longer than VIa (Fig. 219 E); shorter hair on VIa 0.07–0.14 mm, 2–3 × IIIbd. Apical segm. of rostrum about as long as VIa, a little shorter than 2sht. Siphunculi rather short, shorter than longer hair on VIa, with a rather narrow reticulate zone consisting of about three transverse rows of meshes (Figs. 230, 231). Cauda broadly rounded. 1.8–2.6 mm.

Distribution. Not yet found in Scandinavia, but widespread in Europe, south at least to Bulgaria, Ukraine, and Caucasus; very rare in Great Britain; not recorded from N Germany or NW & W Russia; known from Poland and Czechoslovakia.

Biology. The aphids live on leaf stems and tender shoots of *Acer campestre*. Dimorphs have never been observed.

Periphyllus singeri (Börner, 1952)
Figs. 218 F, 219 F, 221 B.

Chaetophoria singeri Börner, 1952: 322. – Survey: 349.

Apterous viviparous female. Dorsum with rather dark cross bars. Apices of antennae dark. Legs rather pale with dark apices. Body and appendages densely covered with fine, mostly rather short, hairs. Processus terminalis about 4.0–4.5 × VIa (Fig. 218 F); VIa with 2–4 long hairs; longest hair on VIa 4–7 times as long as diameter of processus terminalis; shortest hair on VIa 2.5–5.0 × diameter of processus terminalis (Fig. 219 F); ant. segm. III with about 27–29 hairs. Apical segm. of rostrum 1.2–1.5 × VIa, 0.8–1.0 × 2sht., with more than 15 hairs, i. e. with more hairs than in other *P.* spp. (Fig. 221 B).

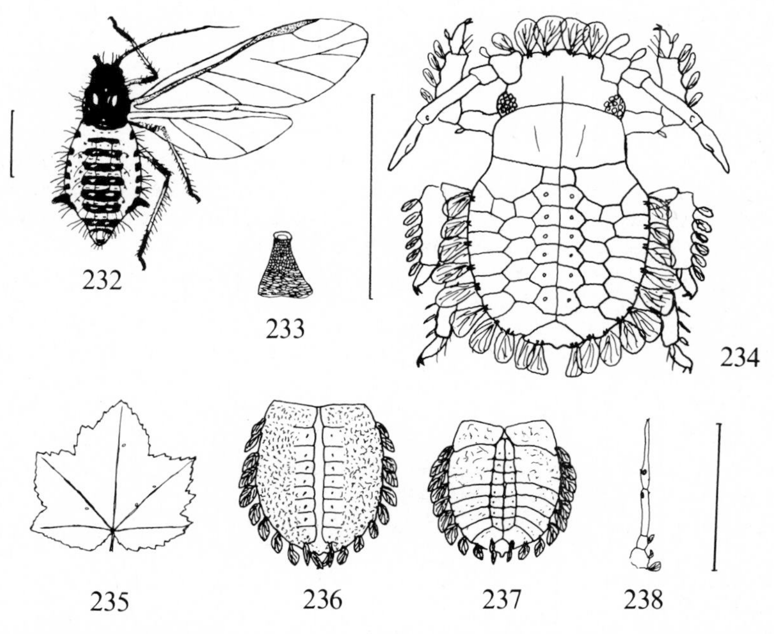

Figs. 232–235. *Periphyllus testudinaceus* (Fernie). – 232: alate viviparous female; 233: siphunculus of al. viv.; 234: dimorph; 235: distribution of dimorphs on leaf of *Acer;* see also Fig. 224.
Figs. 236–238. Dimorphs of *P. hirticornis* (Wlk.) (236 and 238) and *P. californiensis* (Shinji) (237). – 236, 237: meso- and metanotum and abdomen; 238: antenna.
(Scales 1 mm for 232, 0.5 mm for 234 (in the middle) and 236–238 (to the right)). (232 after Essig & Abernathy, 233 and 236–238 after Stroyan, all redrawn, 234 and 235 orig.).

First tarsal segm. with 7 hairs. Siphunculi rather short, with rather narrow reticulate zone, 1–3 narrow transverse rows of meshes which are still finer reticulate themselves. 2.5–3.4 mm.

Alate viviparous female. Dorsal pattern of dark markings as in *acericola.*

Distribution. Not yet found in Scandinavia, but may be overlooked. – The Netherlands, Germany (not N Germany), and Austria, extremely rare; it has been found on less than 10 trees in the world.

Biology. The aphids live on branches, and also on old wounds on the bark of large tree-trunks, of *Acer pseudoplatanus.* They are attended by *Lasius fuliginosus.* Dimorphs are not produced.

123. ***Periphyllus testudinaceus*** (Fernie, 1852)

Plate 4: 5, 6. Figs. 218 D, 219 D, 222 A & B, 232–235.

Phyllophora testudinacea Fernie, 1852: 265.
Phillophorus testudinatus Thornton, 1852: 78.
Periphyllus testudo van der Hoeven, 1863: 1.
? Aphis villosus Hartig, 1841: 369.
Survey: 349.

Apterous viviparous female. Green or dark brown. Ant. segm. I–II and VI, and apices of III–V, dark. Head, pronotum, apices of femora, bases and apices of tibiae (fig. 222 B), tarsi and distal part of siphunculus, dark. Dorsal hairs placed on dark sclerites. Abd. tergites III–V typically each with 2 pairs of spinal hairs, 1 pair of pleural hairs, and 3–5 pairs of marginal hairs. Processus terminalis 2.5–3.7 × VIa (Fig. 218 D), as short as 2.1 × VIa in ovipara-like individual occurring in autumn; longer hair on VIa shorter than VIa, 2.3–2.5 times as long as diameter of processus terminalis, less than 3 times as long as shorter hair on VIa (Fig. 219 D); segm. III with 0–6 rhinaria and 9–13 hairs, as few as 5 hairs in individuals occurring in autumn. Apical segm. of rostrum 1–1.5 × VIa, about 0.75 × 2sht. Hind tibia sometimes with a few scent plaques in individuals from autumn. First tarsal segm. usually with 7, rarely with 5, hairs. Siphunculi about as long as 2 sht., with 5–7 transverse rows of meshes in reticulate zone, 12–13 in alatiform individuals. Cauda at base twice as broad as its length 2.4–3.1 mm.

Fundatrix. Thicker. Processus terminalis only 1.0–1.5 × VIa. Ant. segm. III with 4–5 hairs.

Alate viviparous female. Abdomen with dark marginal sclerites and broad dorsal cross bars darker than pterostigma (Fig. 232). Hind tibiae with dark apices (Fig. 222 A). Siphunculi dark. Pterostigma remarkably pale. Processus terminalis 3.2–4.2 × VIa; longest hair on segm. III 0.11–0.21 mm; longer hair on VIa 0.03–0.05 mm; segm. III with 10–32 rhinaria, number of hairs as in the apterous viviparous female. Cross bar on abd. tergite V with 6 hairs.

Dimorph. Green. Body extremely flat, armoured with sclerotic, spinal and marginal, segmental plates and pleural, intersegmental plates (Fig. 234). With leaf-shaped marginal hairs. Antenna about 0.5 × body, thick, 4-segmented; processus terminalis about as long as basal part of ultimate ant. segm. About 0.7 mm.

Oviparous female. Processus terminalis 3.1–3.2 × VIa. Ant. segm. III with 5–8 hairs. Hind tibia slightly swollen, with many scent plaques all over.

Alate male. Processus terminalis 5.7–5.9 × VIa. Secondary rhinaria on ant. segm. III: about 56–60, IV: 32–33, V: 21–22.

Distribution. Very common and widespread in Denmark and in Sweden, north to Ång.; in Norway recorded from AK, Ry, and HOy; in Finland known from Ab, N, and Ta. – Widespread in Europe, south to Portugal, Spain, and Caucasus, east to Poland and NW Russia; very common in Great Britain, the Netherlands, and Germany. Korea. Tasmania. Widespread in the USA and Canada.

Biology: The species lives on leaves, leaf stems, and young shoots of several species of *Acer,* especially *campestre* and *pseudoplatanus,* occasionally on *Aesculus hippo castanum.* Fundatrices are found in Denmark from April. The dimorphs are the only individuals found from June to September. They occur scattered on the upper – and undersides of leaves (Fig. 235) and are difficult to decern with the naked eye.

Genus *Chaitophorus* Koch, 1854

Chaitophorus Koch, 1854: 1.

Type-species: *Chaitophorus populi* (Linné) of Koch, 1854 = *Aphis populeti* Panzer, 1801.

(The International Commission on Zoological Nomenclature is ruling on the proposal that *Chaitophorus leucomelas* Koch, 1854 be the type species, see Hille Ris Lambers & Stroyan (1975)).

Survey: 137.

Dorsum usually more or less sclerotized; abd. tergites (I–)II–VI fused in most species. Abdomen of alatae with dark segmental cross bars, or sclerites, and more or less developed marginal sclerites. Dorsal cuticle often with nodules, denticles, or fine spinules, sometimes arranged in rows, which may form a reticulate pattern. Processus terminalis longer than VIa (except in fundatrices of some species). Apterae without secondary rhinaria; alatae with secondary rhinaria on ant. segm. III, often also on IV and V, usually arranged in a row. First tarsal segments with 5, 6 or 7 hairs. Empodial hairs rod-shaped. Siphunculi short, stump-shaped, reticulate, at least apically. Cauda knobbed in most species, more or less constricted, sometimes wart-shaped and without constriction in oviparous females; with 4–12 hairs.

The genus is holarctic, with 74 species in the world, 14 species in Scandinavia. They all feed on Salicaceae, some on *Salix,* others on *Populus.*

Important contributions to the taxonomy of the genus have been given by Szelegiewicz (1961) and Stroyan (1977). These papers, together with the original descriptions by Ossiannilsson and others, have been used for the construction of the keys and descriptions below.

Key to species of *Chaitophorus*

Apterous viviparous females (not fundatrices)

1 Hind tibia with scent plaques (Fig. 270). On young shoots and branches of *Populus alba, canescens* and *tremula;* attended by ants. .... 130. *populeti* (Panzer)

– Hind tibia without scent plaques. If on *Populus alba, canescens* or *tremula,* then on leaves and not, or only occasionally, attended by ants. 2

2(1) Siphunculi with transverse lines not forming a real reticulate pattern (Figs. 258, 259). On *Salix repens.* . 128. *parvus* Hille Ris Lambers

– Siphunculi with reticulation. Not on *Salix repens.* 3

3(2) Dorsal hairs with blunt or furcate apices. Apical segm. of rostrum with 2 accessory hairs. 4

– Dorsal hairs with pointed apices, or, if some of them have blunt or furcate apices, then apical segm. of rostrum with more than 2 accessory hairs. 7

4(3) Ant. segm. III with 0–4 hairs, rarely more than 6 on both antennae together. Apical segm. of rostrum 1.0–1.3 × 2sht. On leaves of *Salix.* 5

– Ant. segm. III with 3 or more hairs, rarely less than 8 on both antennae together. On *Salix* or *Populus;* if on *Salix,* then apical segm. of rostrum 0.6–0.9 × 2sht. 6

5(4) Apical segm. of rostrum 1.2–1.3 × 2sht., slender, almost parallel-sided, not sharp-pointed (Fig. 240). Body length (when mounted on slides) 1.8–2.1 × greatest body width. Ant. segm. III with 0–2, usually 1 hair (Fig. 241). On *Salix* spp. with broad leaves. .... 124. *capreae* (Mosley)

– Apical segm. of rostrum 1.0–1.1 × 2sht., elongate triangular, sharp-pointed (Fig. 243). Body length (in slides) 2.1–2.5 × greatest body width. Ant. segm. III with 1–4, usually 2–3 hairs (Fig. 244). On *Salix* spp. with narrow leaves. . 125. *horii beuthani* (Börner)

6(4) First tarsal segm. usually with 5 hairs, less often with 4, 6 or 7 (evaluate from sample of several specimens). On *Populus.* .... 131. *populialbae* (Boyer de Fonscolombe)

– First tarsal segm. usually with 7 hairs, less often with 5, 6 or 8 (valuate from sample of several specimens). On *Salix.* 14

7(3) Abd. tergites all free. .... 132. *ramicola* (Börner)

– Abd. tergites (I–)II–VI fused, not separated by membranous intersegmental border-zones. (Fundatrices of *lapponum* Ossiannilsson, which can be found in mid-summer, frequently have free tergites, see Fig. 250). 8

8(7) Dorsal cuticle with fine spinules in wavy rows, sometimes forming a reticulate pattern. First tarsal segm. usually with 5 hairs (evaluate from sample of several specimens). On *Salix*. 9

– Dorsal cuticle with denticles or nodules not placed in rows or forming a reticulate pattern, or not sculptured at all. First tarsal segm. usually with 6–7 hairs (evaluate from sample of several specimens). On *Salix* or *Populus*. 11

9(8) Apical segm. of rostrum 1.09–1.5 × 2sht. .... 133. *salicti* (Schrank)

– Apical segm. of rostrum 1.07 × 2sht. or shorter. 10

10(9) Processus terminalis longer than ant. segm. III. .... 134. *salijaponicus niger* Mordvilko

– Processus terminalis shorter than ant. segm. III. .... 126. *lapponum* Ossiannilsson

11(8) Longer antennal hairs rather evenly distributed round circumference of segm. III–V. Processus terminalis twice as long as VIa or shorter. On young shoots and twigs of *Salix* spp.; always attended by ants. .... 137. *vitellinae* (Schrank)

– Longer antennal hairs directed predominantly anterad (on inner face of segments). Processus terminalis more than twice as long as VIa. On *Populus* or on leaves of *Salix;* not attended by ants, or rarely so. 2

12(11) Dorsal cuticle rather densely sculptured with denticular spinules. Dorsum blackish. Abd. tergite I more or less completely fused with tergites II–VI. On leaves of *Populus tremula*. 135. *tremulae* Koch

– Dorsal cuticle with flattish or roundish nodules, usually only sparsely sculptured, in some cases not sculptured. Dorsum pale or blackish; if blackish, then abd. tergite I not solidly fused with tergites II–VI. Not on *Populus tremula*. 13

13(12) Apical segm. of rostrum 0.9–1.0 × 2sht., with 4–9 accessory hairs. Ant. segm. III with 9–18 hairs. On *Populus*. .. 127. *leucomelas* Koch

– Apical segm. of rostrum 0.6–0.9 × 2sht., with 2–4 accessory hairs. Ant. segm. III with 6–11 hairs. On *Salix*. 14

14(6,13) Longest hair on VIa shorter than 0.5 × VIa. Longest hair on first tarsal segm. shorter than 0.5 × 2sht. .... 136. *truncata* (Hausmann)

– Longest hair on VIa longer than 0.5 × VIa. Longest hair on first tarsal segm. longer than 0.5 × 2sht. .... 129. *pentandrinus* Ossiannilsson

Alate viviparous females

1 Hind tibiae with scent plaques. .... 130. *populeti* (Panzer)

– Hind tibiae without scent plaques. 2

2 (1) First tarsal segm. typically with 5 hairs (evaluate from sample of several specimens). 3

– First tarsal segm. typically with 6 or 7 hairs. 10
3 (2) Apical segm. of rostrum shorter than hind tarsus (claws not included). 4
– Apical segm. of rostrum as long as hind tarsus or longer. 8
4 (3) Processus terminalis shorter than 0.75 × ant. segm. III.
126. *lapponum* Ossiannilsson
– Processus terminalis longer than 0.75 × ant. segm. III. 5
5 (4) Abdomen with narrow dorsal cross bars. Ant. segm. III with 3–6 rhinaria. Processus terminalis about 2.25 × VIa. On *Salix repens*. .... 128. *parvus* Hille Ris Lambers
– Abdomen with broad dorsal cross bars, which may be more or less fused. If ant. segm. III has less than 7 secondary rhinaria, then cross bars on abd. tergites III–VI are fused (Fig. 246). Processus terminalis usually longer than 2.5 × VIa. Not on *Salix repens*. 6
6 (5) Longest hair on ant. segm. III longer than 4 × IIIbd.
134. *salijaponicus niger* Mordvilko
– Longest hair on ant. segm. III shorter than 2.5 × IIIbd. 7
7 (6) Marginal sclerites somewhat paler than mid-dorsal patch formed by fusion of cross bars on abd. tergites III–VI. Abd. tergite VIII with less than 12 hairs. Ant. segm. III with 7–26 rhinaria and 3–12 hairs; segm. IV with 1–5 rhinaria. .... 131. *populialbae* (Boyer de Fonscolombe)
– Marginal sclerites not noticeably paler than mid-dorsal patch formed by fusion of abd. tergites III–VI. Abd. tergite VIII with more than 12 hairs. Ant. segm. III with 4–7 rhinaria and 0–4 hairs (rarely more than 6 hairs on both antennae together); segm. IV with 0–4 rhinaria.
125. *horii beuthani* (Börner)
8 (3) Longest hair on ant. segm. III longer than 4 × IIIbd. Apical segm. of rostrum with (3–)4(–5) accessory hairs. .... 133. *salicti* (Schrank)
– Longest hair on ant. segm. III shorter than 2.5 × IIIbd. Apical segm. of rostrum with 1–2 accessory hairs. 9
9 (8) Dorsal sclerites not forming anything like a mid-dorsal rectangular patch. Marginal sclerites not pigmented. Ant. segm. III with 3–7 rhinaria. .... 124. *capreae* (Mosley)
– With a dark rectangular dorsal patch formed by fusion of cross bars on abd. tergites III–VI. Marginal sclerites pigmented, but somewhat paler than mid-dorsal patch. Ant. segm. III with 7–26 rhinaria. .... 131. *populialbae* (Boyer de Fonscolombe)
10 (2) Apical segm. of rostrum longer than 1.0 × 2sht. On *Salix*. Longer antennal hairs rather evenly distributed round circumference of segm. III–V. Cauda only little constricted. 11

– Apical segm. of rostrum 1.0 × 2sht. or shorter. On *Salix* or *Populus;* if on *Salix,* then apical segm. of rostrum 0.6–0.9 × 2sht. Longer antennal hairs placed mainly on inner face of segm. III–V. Cauda distinctly constricted. 12

11 (10) Processus terminalis 1.2–1.6 × VIa, 0.5–0.6 × ant. segm. III. 137. *vitellinae* (Schrank)

– Processus terminalis longer than 1.7 × VIa, almost as long as ant. segm. III. .... 132. *ramicola* (Börner)

12 (10) Antenna shorter than 0.65 × body. Longest hair on VIa considerably longer than 0.5 × VIa. .... 129. *pentandrinus* Ossiannilsson

– Antenna longer than 0.65 × body. Longest hair on VIa about as long as 0.5 × VIa, or shorter. 13

13 (12) Ant. segm. III with more than 25 hairs on both antennae together. .... 127. *leucomelas* Koch

– Ant. segm. III with less than 25 hairs on both antennae together. 14

14 (13) Dorsal pigmented cross bars on abd. tergites broad, solid, and nearly fusing into a solid patch on tergites III–VI in many specimens. Apical segm. of rostrum with (3–) 4–6 accessory hairs. Ant. segm. III with 19–24 hairs on both antennae together. Secondary rhinaria rather irregularly arranged on ant. segm. III. .... 135. *tremulae* Koch

– Dorsal pigmented cross bars on abd. tergites much broken up and not forming anything like a solid mid-dorsal patch. Apical segm. of rostrum with 2 (–3) accessory hairs. Ant. segm. III with 16–19 hairs on both antennae together. Secondary rhinaria arranged in a single row, or in a partly double row on ant. segm. III. .... 136. *truncatus* (Hausmann)

124. ***Chaitophorus capreae*** (Mosley, 1841)
Figs. 239–241.

Cinara capreae Mosley, 1841: 748. – Survey: 138.

Apterous viviparous female. Whitish. Apices of ant. segm. rather dark. Body (Fig. 239) broader than in *horii beuthani.* Abd. tergites I–VI fused. Dorsal cuticle without microsculpture, not pigmented. Dorsal hairs furcate; abd. segm. III with 12–24 such hairs, usually in a single transverse row; each abd. segm. with six hairs longer than the others. Antenna about 0.6 × body; processus terminalis 2.0–3.3 × VIa, distinctly longer than segm. III; segm. III with 0–2 (usually 1) hairs, the longer one 0.5–1.0 × IIIbd. (Fig. 241). Apical segm. of rostrum 1.2–1.3 × 2sht., slender, with 2 accessory hairs (Fig. 240). First tarsal segm. with 5 hairs. Abd. tergite VIII with 10–15 hairs. 0.8–1.9 mm.

Fundatrix. Body broader. Antenna about 0.4 × body; processus terminalis about 1.3–2 × VIa.

Alate viviparous female. Head and thorax brown. Abdomen only slightly sclerotic, without dark dorsal markings, or with a much broken pattern of sclerites. Antennae brown. Secondary rhinaria on ant. segm. III: 3–7, IV: 0–1, V: 0; ant. segm. III with 1–3 hairs, the longest one 0.9–1.3 × IIIbd. Abd. tergite VIII with about 10–11 hairs.

Oviparous female. Dirty yellow. Antennae and legs dark. Hind tibia only slightly thickened, with rather few scent plaques in the middle.

Apterous male. Yellow, with brown head and brown longitudinal median stripe on dorsum. Antenna a little shorter than body; secondary rhinaria on segm. III: 5–8, IV: 4–6, V: 2–4.

Distribution. In Denmark common and widespread; in Sweden common in the south, north to Med.; in Norway recorded from On and Nsy; not in Finland. – In Europe and Asia, south to Portugal, Spain, and Transcaucasia, east to W Siberia and C Asia; very common in Great Britain; not rare in N Germany; known from the Baltic region of Poland, but not from NW & W Russia.

Biology. The aphids live on the undersides of leaves of *Salix* spp. with broad leaves, *S. caprea, aurita, cinerea,* and *lapponum.* They are not visited by ants.

125. ***Chaitophorus horii*** Takahashi, 1939

Chaitophorus horii Takahashi, 1939: 122. – Survey: 139.

The species is subdivided into two subspecies, viz. *h. horii* from the Far East and *h. beuthani* from Europe and N Asia.

Chaitophorus horii beuthani (Börner, 1950)
Plate 4: 8. Figs. 242–246.

Tranaphis beuthani Börner, 1950: 3. – Survey: 139.

Apterous viviparous female. Whitish. Ant. segm. VI and siphunculi a little darker than other parts of body. Body elongate, oval (Fig. 242). Abd. tergites I–VI fused. Dorsal cuticle without microsculpture, not pigmented. Dorsal hairs thick, furcate (Fig. 245); abd. segm. III with 25–45 such hairs in a distinct double row; each abd. segm. with six hairs longer than the rest. Antenna 0.4–0.6 × body; processus terminalis 2.4–3.7 × VIa, distinctly longer than segm. III; segm. III with 1–4, usually 2–3, such hairs, the longest one 1.0–2.1 × IIIbd. (Fig. 244). Apical segm. of rostrum 1.0–1.1 × 2sht., pointed, with 2 accessory hairs (Fig. 243). First tarsal segm. with 5(–6) hairs. Abd. tergite VIII with 14–26 hairs. 1.0–1.8 mm.

Fundatrix. Body broader. Antenna about 0.4 × body; processus terminalis 1.6 × VIa, or shorter.

Alate viviparous female. Head and thorax yellowish pigmented. Abdomen with dark marginal sclerites and dorsal cross bars; those on tergites III–VI tend to form a solid rectangular mid-dorsal patch (Fig. 246). Antennae brownish. Legs and siphunculi brown. Secondary rhinaria on ant segm. III: 4–7, IV: 0–2, V: 0; ant. segm. III with 2–4

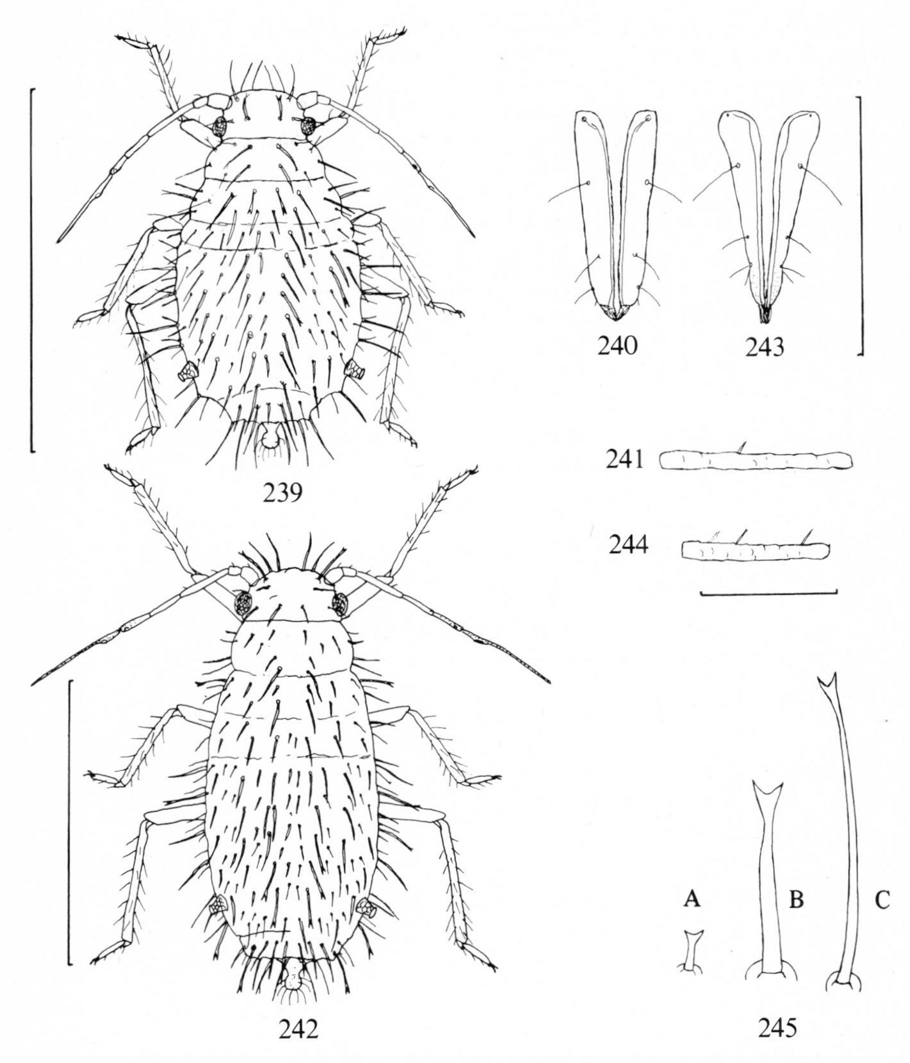

Figs. 239–241. *Chaitophorus capreae* (Mosl.), apt. viv. – 239: habitus; 240: apical segm. of rostrum; 241: ant. segm. III.

Figs. 242–245. *C. horii* subsp. *beuthani* (Börner), apt. viv. – 242: habitus; 243: apical segm. of rostrum; 244: ant. segm. III (the dotted hair is often absent); 245: furcate abdominal hairs, A: short pleural, B: long pleural, C: long marginal.

(Scales 1 mm for 239 and 242, 0.1 mm for 240 and 243, 0.1 mm for 241 and 244).

hairs, the longest one 1.5–2.1 × IIIbd. Abd. tergite VIII with about 17–19 hairs.

Oviparous female. Brownish yellow. Antennae and legs brown. Body spindle-shaped. Dorsum membranous. Processus terminalis 2.3–2.8 × VIa. Hind tibia not swollen, with rather few scent plaques. Cauda not constricted. Rather large, about 2 mm.

Apterous male. Head black. Thorax and abdomen yellow, with dark dorsal cross bars on abd. tergites II–VI. Antennae, legs, and siphunculi, dark. Body slender. Antenna about 0.7 × body; secondary rhinaria on segm. III: 5–12, IV: 5–12, V: 3–7.

Distribution. In Denmark rather common and widespread; in Sweden found in Sk. and Upl.; not in Norway or Finland. – Common in Great Britain and N Germany; in Europe south at least to Austria and Czechoslovakia, east to Poland.

Biology. The aphids live on the undersides of leaves of *Salix viminalis* and other *S.* spp. with lanceolate or linear leaves. They are not visited by ants.

126. ***Chaitophorus lapponum*** Ossiannilsson, 1959
Figs. 249–253.

Chaitophorus lapponum Ossiannilsson, 1959: 1. – Survey: 139.

Apterous viviparous female. Black. Siphunculus surrounded by pale membranous area. Abd. tergites II–VI fused. Dorsal cuticle with reticulate microsculpture (Fig. 251). Dorsal hairs long, pointed. Antenna about 0.5 × body; processus terminalis 1.8–2.4 × VIa, shorter than segm. III; segm. III with about 8 hairs; most of the longer, erect, antennal hairs are placed on inner surface of segments, 1–3 fairly long antennal hairs may be present on outer surface (Fig. 253); longest antennal hairs as long as VIa; the two hairs on VIa 1–1.5 × largest diameter of VIa (shorter than in the related species *ramicola*). Apical segm. of rostrum 0.7–0.9 × 2sht., with 4–6 accessory hairs. First tarsal segm. typically with 5 hairs (occasionally with 6–7, most frequently so on hind tarsi). Cauda knobbed. 1.5–1.8 mm.

Fundatrix. Abd. tergites II–VI more or less fused, sometimes apparently free (Fig. 250). Antenna shorter than 0.5 × body; processus terminalis 1.0–1.5 × VIa. Cauda more or less distinctly knobbed. Rather large, 2.0–2.3 mm.

Alate viviparous female. Abdomen with dark dorsal cross bars and large marginal sclerites (Fig. 249). Secondary rhinaria on ant. segm. III: 5–10, IV: 0–4, V: 0 (Fig. 252).

Distribution. Known only from the northern part of Sweden, from Dlr. north to Lu.Lpm, and in Finland from Ok and ObN.

Biology. The host plants are *Salix lapponum, glauca, nigricans,* and *phylicifolia.*

Note. Szelegiewicz (1968) regarded *mordvilkoi* Mamontova, 1960 (known from Ukraine, Rumania, and S Poland), as a subspecies of *lapponum.* This has not been accepted by Eastop & Hille Ris Lambers (1976). *C. mordvilkoi* is, according to the description given by Szelegiewicz (1961), different from *lapponum* in several characters.

The apical segm. of rostrum has only 2 accessory hairs, processus terminalis is as long as, or longer than, ant. segm. III, and the alata has a lower number of secondary rhinaria.

127. ***Chaitophorus leucomelas*** Koch, 1854
Plate 4: 7. Fig. 254.

Chaitophorus leucomelas Koch, 1854: 4.
Chaitophorus versicolor Koch, 1854: 10.
Survey: 139.

Apterous viviparous female. Colour variable, green or yellow, typically with two darker longitudinal pleural stripes, which may fuse across midline on some of the posterior abd. segments, or may be reduced, or even absent. Head and thorax darker. Apices of

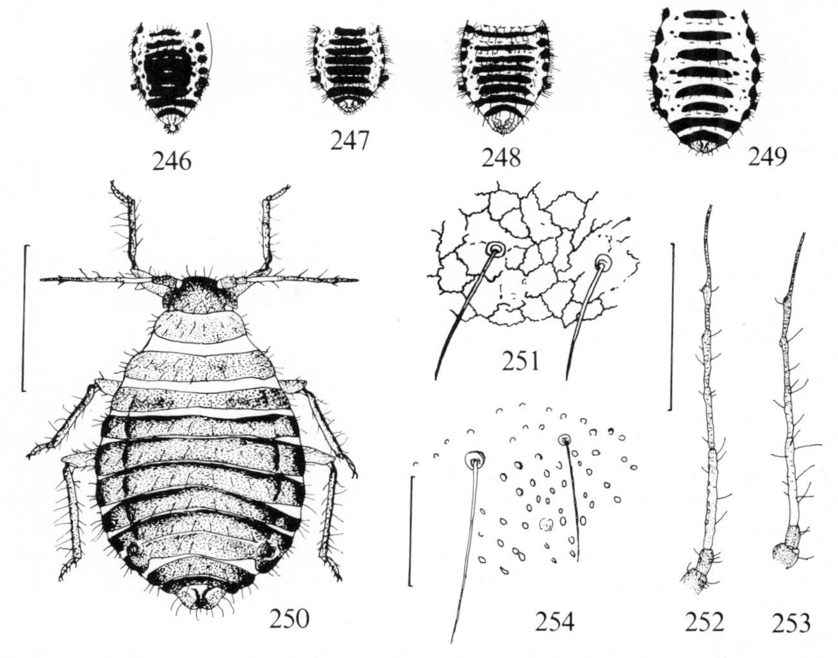

Figs. 246–249. Abdomen of al. viv. of *Chaitophorus* spp. – 246: *horii* subsp. *beuthani* (Börn.); 247: *salijaponicus* subsp. *niger* Mordv.; 248: *vitellinae* (Schrk.); 249: *lapponum* Ossiann. (246–248 after Stroyan, redrawn; 249 orig.).
Figs. 250–253. *Chaitophorus lapponum* Ossiann. – 250: fundatrix (paratype), abd. tergites free; 251: part of dorsal cuticle of abd. segm. IV of same; 252: antenna of al. viv.; 253: antenna of apt. viv. (Scales 1 mm for 250, 0.5 mm for 252 and 253). (252 and 253 after Ossiannilsson, redrawn).
Fig. 254. *C. leucomelas* Koch, part of dorsal cuticle of abd. segm. V of apt. viv. (Scale 0.1 mm).

ant. segm., femora, tarsi, and siphunculi, dark. Abd. tergites II–VI fused. Dorsal cuticle with nodules (Fig. 254), conspicuous only on pigmented areas. Dorsal hairs long, 3–6.5 × IIIbd., pointed. Antenna longer than 0.5 × body; processus terminalis 2.7–3.3 × VIa, alout as long as segm. III; segm. III with 9–18 hairs, the longest one about 3 × IIIbd. Apical segm. of rostrum 0.9–1.0 × 2sht., with 4–9 (usually 7–8) accessory hairs. First tarsal segm. with 7 hairs. Cauda distinctly knobbed. 1.2–2.0 mm.

Fundatrix. Grass green or yellowish green. Abdomen without nodules. Dorsal hairs thick, often blunt, only 3–4 × IIIbd. Antenna 0.5 × body or shorter; processus terminalis about 1.8–2.1 × VIa.

Alate viviparous female. Head, thorax, antennae (except base of segm. III), and siphunculi, black. Abdomen green or yellowish green with dark dorsal cross bars, and equally dark marginal sclerites. Cross bars on tergites III–VI broad, solid, nearly fused into a mid-dorsal patch. Antenna 0.7 × body; processus terminalis 2.8–4.1 × VIa; secondary rhinaria on segm. III: 9–18, IV: 0–1, V: 0; segm. III with 26–34 hairs on both antennae together.

Oviparous female. Yellow or dirty yellow. Head, pronotum, sclerites of mesonotum, and (VII–) VIII abd. tergites light brown. Siphunculi pale. Processus terminalis about 3×VIa. Hind tibia swollen, with numerous scent plaques. Cauda dark, without constriction. Rather large, 2.3–2.6 mm.

Alate male. Head and thorax black. Abdomen light brown with darker marginal sclerites and dorsal cross bars. Siphunculi and cauda dark. Antennae dark, almost as long as body; processus terminalis about 3 × VIa; secondary rhinaria on segm. III: 29–52, IV: 23–37, V: 6–27.

Distribution. In Denmark not very common, but widespread; in Sweden known from Sk. north to Upl.; in Norway known from Bø; not in Finland. – Widespread in Europe and Asia, south to Portugal, Spain, Yugoslavia, Bulgaria, Turkey, and Iran, east to Siberia, Mongolia, and C Asia; common in Great Britain; rare in N Germany; recorded from Poland and Russia. Widespread in the USA and Canada.

Biology. The aphids live on the undersides of leaves, and on young shoots, of *Populus nigra* and *italica,* often in empty galls caused by aphids (*Pemphigus* spp. a. o.), or other insects. They are visited by ants in spring.

128. ***Chaitophorus parvus*** Hille Ris Lambers, 1935
Figs. 255–260.

Chaitophorus parvus Hille Ris Lambers, 1935: 53. – Survey: 141.

Apterous viviparous female. Black; greenish as a nymph. Abd. tergites II–VI fused (Fig. 255). Dorsal cuticle with wavy rows of fine denticles (Fig. 256), which sometimes form a reticulate pattern on thorax (Fig. 257). Dorsal hairs long, up to 8 × IIIbd., pointed. Antenna about 0.5 × body; processus terminalis 1.8–2.1 × VIa, a little longer than segm. III; segm. III with 4–6 hairs, the longest one about 2.5–3.5 × IIIbd. Apical segm. of rostrum about as long as 2sht., with 4 accessory hairs. First tarsal segm. with 5 hairs.

Siphunculus (Fig. 259) with sculpture of transverse, almost parallel, thick, black lines, touching each other in some points, but not forming a proper reticulate pattern as in other *C.* spp. Cauda only slightly constricted (Fig. 260). 1.2–1.8 mm.

Alate viviparous female. Abdomen with narrow cross bars. Secondary rhinaria on ant. segm. III: 3–6, IV: 0–2, V: 0.

Distribution. In Denmark known from NEJ (Læsø and Blokhus); in Sweden found in Sk. (Skanör) and Boh. (Skaftö); not in Norway or Finland. – Described from the northern part of Poland; not yet found outside Poland, Denmark, and Sweden.

Biology. The aphids live on *Salix repens* (= *rosmarinifolia*), mostly on the undersides of leaves. The colonies are visited by ants. The species has in Denmark been found in dunes only.

129. ***Chaitophorus pentandrinus*** Ossiannilsson, 1959
Figs. 264–267.

Chaitophorus pentandrinus Ossiannilsson, 1959: 3. – Survey: 141.

Apterous viviparous female. Pale green or yellow. Dorsal cuticle with numerous nodules. Dorsal hairs sword-shaped with pointed, blunt, or furcate apices; marginal

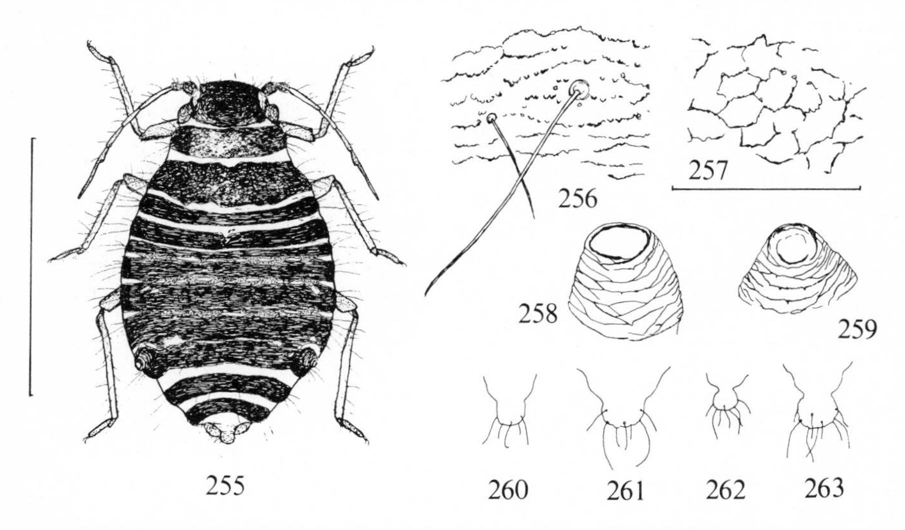

Figs. 255–259. *Chaitophorus parvus* H. R. L. – 255: apt. viv.; 256: part of dorsal cuticle of abd. segm. III of apt. viv.; 257: part of dorsal cuticle of mesothorax of apt. viv.; 258: siphunculus of al. viv.; 259: siphunculus of apt. viv. (Scales 1 mm for 255, 0.1 mm for 256).
Figs. 260–263. Cauda of apt. viv. of *Chaitophorus* spp. – 260: *parvus* H. R. L.; 261: *ramicola* (Börn.); 262: *salicti* (Schrk.); 263: *salijaponicus* subsp. *niger* Mordv. (After Szelegiewicz, redrawn in outline).

hairs often long with fine-pointed apices (Fig. 265, left). Antenna about 0.5 × body or shorter; processus terminalis 2.3–3.0 × VIa, 0.9–1.4 × segm. III; antennal hairs pointed, the longest about as long as ant. segm. IV; ant. segm. III with about 6–7 hairs; VIa with 3 fairly long hairs, the longest one almost as long as length of VIa proximally of the rhinarium (Fig. 267). Apical segm. of rostrum 0.8–0.9 × 2sht., with 2–4 accessory hairs. First tarsal segm. with 6 or 7 hairs. Cauda distinctly knobbed. 1.4–2.3 mm.

Alate viviparous female. Head, thorax, ant. segm. I–II, and siphunculi, dark. Abdomen with dark marginal sclerites and dorsal cross bars. Cross bars on tergites III–VI often more or less fused. Dorsal hairs pointed. Antenna (Fig. 266) a little longer than 0.5 × body; processus terminalis 2.7–3.2 × VIa; secondary rhinaria on segm. III: 9–14, IV: 1–5, V: 0–2.

Apterous male. Black or brown. Dorsal sclerites on abd. segm. III–VI more or less fused into a patch. Antenna 0.7–0.8 × body; secondary rhinaria on segm. III: 12–21, IV: 6–9, V: 6–9. Rather small, 1.2–1.3 mm.

Distribution. Known only from Ög. (Rystad) and Upl. (Uppsala) in Sweden.

Biology. The host is *Salix pentandra.*

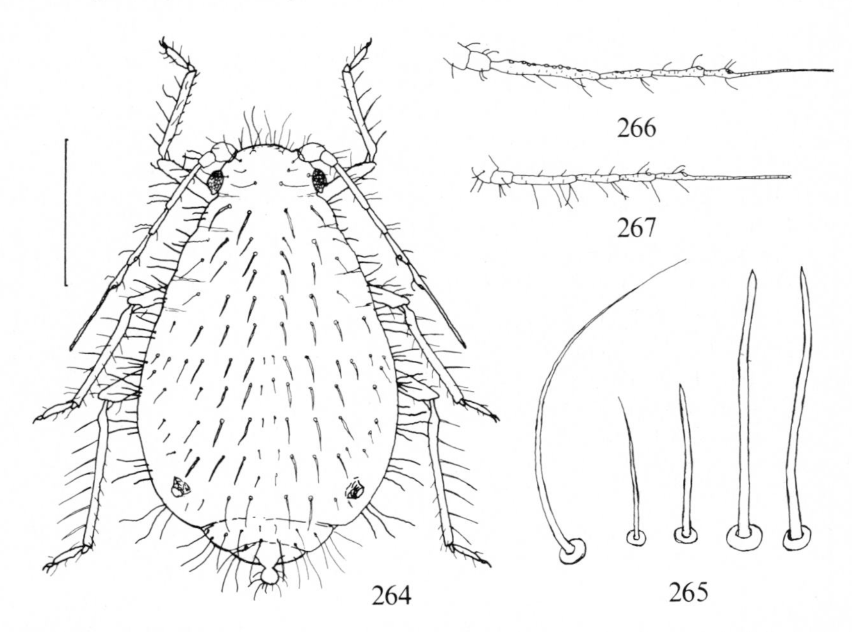

Figs. 264–267. *Chaitophorus pentandrinus* Ossiann. – 264: apt. viv. (paratype); 265: types of abdominal hairs of apt. viv., marginal hair to the left, four dorsal hairs to the right; 266: antenna of al. viv.; 267: antenna of apt. viv. (Scale 0.5 mm for 264, 266 and 267). (266 and 267 after Ossian-nilsson, redrawn).

130. ***Chaitophorus populeti*** (Panzer, 1801)
Figs. 268–271.

Aphis populeti Panzer, 1801: 18.
Chaitophorus populi (Linné) of Koch, 1854: 12.
Chaitophorus betulinus van der Goot, 1912: 276. Survey: 141.

Apterous viviparous female. Shining dark green to black, with paler longitudinal median stripe on thorax and anterior part of abdomen. Antennae (except basal part of flagellum), legs, and siphunculi, dark. Abd. tergites I–VI fused. Dorsal cuticle without microsculpture. Dorsal hairs pointed, or short and thick, with blunt or furcate apices, 2.0–3.7 × IIIbd. Antenna longer than 0.5 × body; processus terminalis about 2 × VIa, considerably shorter than segm. III; segm. III with 10–28 erect hairs, about 2 × IIIbd. or longer (Fig. 268); the longer antennal hairs often arranged in about equal number on inner and outer surface of segments (as in *vitellinae;* most other *C.* spp. with longer hairs have these exclusively, or predominantly, on inner surface of segments). Apical segm. of rostrum 1.1–1.4 × 2sht., with 5 to 7 accessory hairs. First tarsal segm. with (6–)7(–9) hairs. Hind tibia with 1–20 scent plaques (Fig. 270), rarely without scent plaques on one of the tibiae. Cauda distinctly constricted. 2.0–2.3 mm.

Fundatrix. Antenna 0.4–0.5 × body; processus terminalis as long as VIa. Hind tibia with a few scent plaques. Larger than apterae of later generations, nearly 3 mm long.

Alate viviparous female. Black. Abdomen with marginal sclerites and broad dorsal cross bars. Secondary rhinaria on ant. segm. III: 10–27, IV: 0–8, V: 0 (Fig. 269). Hind tibia with a few scent plaques. Wing veins brown-shadowed.

Oviparous female. Light brown. Antennae and legs black. Siphunculi pale. Cauda wart-shaped. Hind tibia slightly swollen, with numerous scent plaques.

Male. Alate or apterous. Black. Secondary rhinaria on ant. segm. III: 30–50, IV: 18–36, V: 10–16. Hind tibia with a few scent plaques. Rather small, 2 mm or shorter.

Distribution. In Denmark common in some years, widespread; in Sweden common and widespread all over the country, from Sk. in the south to T. Lpm. in the north; in Norway found in several districts in the southern part of the country, north to MRy; in Finland common and widespread, north to ObN and Ks. – All over Europe; uncommon in Great Britain; common in N Germany; known from the Baltic region of Poland and NW & W Russia. Asia: Turkey, Israel, Iran, Siberia, Caucasus, Kazakhstan, C Asia, Korea. N Africa: Morocco, Egypt.

Biology. The aphids live in colonies on young shoots and branches of *Populus alba, tremula,* and *canescens.* The colonies are visited by ants.

Note. A second European species with scent plaques on hind tibiae of viviparous females is *nassonowi* Mordvilko on *Populus italica* and *nigra.* Its chief distinguishing character is the short processus terminalis, not only in fundatrix, but also in viviparae of later generations only being 1.0–1.2 × VIa (Szelegiewicz 1961). It lives in C, S & E Europe, probably not in Fennoscandia or Denmark.

131. ***Chaitophorus populialbae*** (Boyer de Fonscolombe, 1841)
Figs. 275–277.

Aphis populialbae Boyer de Fonscolombe, 1841: 187.
Chaitophorus albus Mordvilko, 1901: 410.
Survey: 141.

Apterous viviparous female. Whitish or pale green, often with small green spots. Head, apices of ant. segments, and tarsi, brownish. Abd. tergites II–VI fused (Fig. 275). Dorsal cuticle with indistinct reticulation on thorax, rows of fine points on abdomen. Many of the dorsal and antennal hairs are thick, with expanded or furcate apices (Fig. 277), on dorsum 2.5–5.5 × IIIbd. Antenna 0.6–0.9 × body; processus terminalis 2.6–3.3 × VIa,

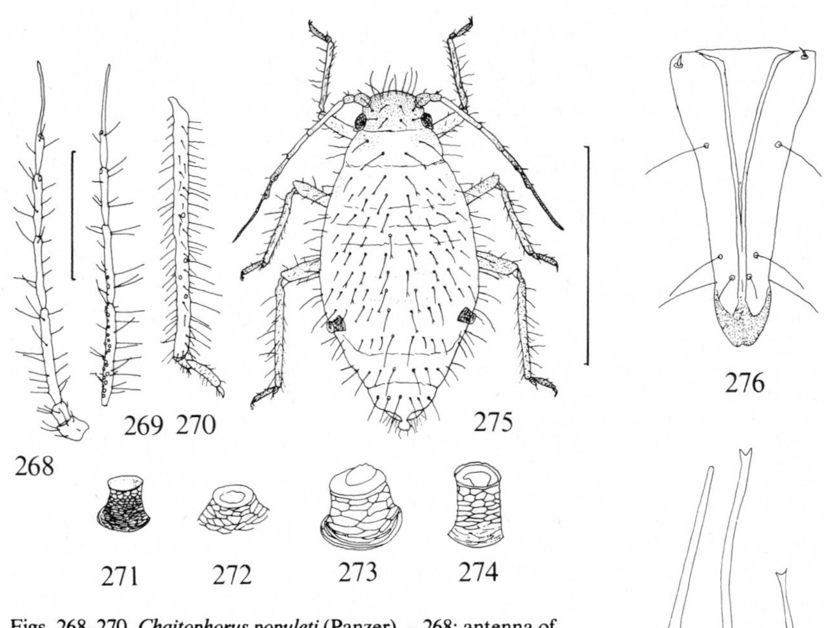

Figs. 268–270. *Chaitophorus populeti* (Panzer). – 268: antenna of apt. viv.; 269: antennal flagellum of al. viv.; 270: hind tibia and tarsus of apt. viv. (with scent plaques). (Scale 0.5 mm). (After Stroyan, redrawn).
Figs. 271–274. Siphunculi of *Chaitophorus* spp., apt. viv. – 271: *populeti* (Panz.); 272: *ramicola* (Börn.); 273: *salijaponicus* subsp. *niger* Mordv.; 274: *truncatus* (Hausm.). (271, 273, 274 after Stroyan, 272 after Szelegiewicz, redrawn).
Figs. 275–277. *C. populialbae* (B. d. F.), apt. viv. – 275: habitus; 276: apical segm. of rostrum; 277: dorsal hairs. (Scale 1 mm for 275).

about as long as segm. III or longer; segm. III with 3–12 hairs, the longest one 0.9–2.0 × IIIbd. Apical segm. of rostrum 0.9–1.2 × 2sht., with 2 accessory hairs (Fig. 276). First tarsal segm. with 5 hairs. Abd. tergite VIII with 6–9 hairs. Cauda distinctly constricted 1.1–2.0 mm.

Fundatrix. Light green. Apices of ant. segments darker. Dorsal cuticle with distinct reticulation. Antenna shorter than 0.5 × body; processus terminalis only a little longer than VIa. Rather large, 2.0–2.3 mm.

Alate viviparous female. Head, thorax, and antennae, black. Abdomen green or yellow, with dorsal sclerites, or cross bars, and a rectangular dorsal patch formed by fusion of cross bars on tergites III–VI, dark; marginal sclerites lightly pigmented, somewhat paler than mid-dorsal patch. Dorsal and antennal hairs pointed. Secondary rhinaria on ant. segm. III: 7–26, IV: 0–5, V: 0–2; longest hair on segm. III 1.3–2.4 × IIIbd. Abd. tergite VIII with 7–11 hairs.

Oviparous female. Yellowish green. Antennae and legs dark. Antenna about 0.5 × body. Hind tibia slightly thickened, with numerous scent plaques. Otherwise rather similar to the apterous viviparous female. Body length as in fundatrix.

Apterous male. Almost black. Antennae, legs, siphunculi, and cauda rather dark. Abdomen with marginal sclerites and dorsal cross bars. Secondary rhinaria on ant. segm. III: 11–53, IV: 17–35, V: 10–19. About 1.5 mm. – Alate males have also been found (even in the same sample). About 2 mm.

Distribution. In Denmark apparently rare, known only from NWJ (Thy) and NEZ (Tåstrup, in trap); in Sweden found in Sk., Upl., Lu.Lpm., and T.Lpm.; in Norway known from AK (trap); not in Finland. – All over Europe, including Great Britain (local), N Germany (rare), Poland, and Russia; south to the Mediterranean Sea. Asia: Turkey, W Siberia, C Asia, Mongolia. Africa: Egypt. N America: widespread in the USA and Canada.

Biology. The aphids occur on the undersides of leaves of *Populus alba, tremula,* and *canescens.* The colonies are sometimes visited by ants.

132. ***Chaitophorus ramicola*** (Börner, 1949)
Figs. 261, 272.

Promicrella ramicola Börner, 1949: 54. – Survey: 142.

Apterous viviparous female. Dark greyish olive. Antennae and legs dark. Siphunculi light brown. Abd. tergites free (not fused). Dorsum dark pigmented, with membranous areas around siphunculi. Dorsal cuticle densely sculptured with rows of fine nodules and pointed hairs; hairs about 4 × IIIbd. Antenna a longer little longer than 0.5 × body; processus terminalis 1.7–2.0 × VIa, as long as or a little longer than segm. III; segm. III with 7–11 hairs, the longest one about 2 × diameter of segment; hairs on VIa longer than 1.5 × largest diameter of segm. VI. Apical segm. of rostrum a little longer than 2sht., with 6 accessory hairs. First tarsal segm. with 6 hairs. Cauda only slightly constricted (Fig. 261). 1.5–1.8 mm.

Fundatrix. Darker, sometimes greyish brown. Antenna shorter than 0.5 × body; processus terminalis about 1.5 × VIa. Rather large, 2.0–2.2 mm.

Alate viviparous female. Abdomen with dark dorsal cross bars. Secondary rhinaria on ant. segm. III: 6–8, IV: 0, V: 0.

Oviparous female. Rather elongate. Antenna shorter than 0.5 × body; processus terminalis about 1.5 × VIa. Hind tibia thickened, with about 5–10 scent plaques. Rather large, 2.4 mm.

Apterous male. Very dark. Antenna about 0.65 × body; secondary rhinaria on segm. III: 7–11, IV: 6–10, V: 1–2. Otherwise much like the apterous viviparous female.

Distribution. In Sweden known from Upl. (Vaksala); not in Denmark, Norway, and Finland. – The Netherlands, Germany (not N Germany), Poland, Czechoslovakia, Austria, NW Russia, Ukraine; apparently a rare species.

Biology. The species lives in small colonies on *Salix cinerea* and *caprea*. The colonies are found on twigs, especially near the place where they branch out, and close to the ground. Ants build earth galleries around the aphids.

Note. The species is very similar to *lapponum*. The number of hairs on first tarsal segments is a variable character, and the abdominal tergites II–VI may be more or less free in both species. The longer hairs on VIa in *ramicola* may be used as a distinguishing character according to Ossiannilsson (see description of *lapponum*).

133. ***Chaitophorus salicti*** (Schrank, 1801)
Fig. 262.

Aphis salicti Schrank, 1801: 103. – Survey: 142.

Apterous viviparous female. Usually black with pale longitudinal median band; some specimens in late summer yellowish white with reddish brown or greyish black dorsal markings. Antennae or apices of ant. segments, at least, and siphunculi, dark. Abd. tergites II–VI fused. Dorsal cuticle with wavy, dense rows of fine denticles, distinct only in dark individuals. Dorsal hairs long, about 7–8 × IIIbd., pointed. Antenna longer than 0.5 × body; processus terminalis about 2–3 × VIa, usually a little shorter than segm. III; segm. III with 7–12 hairs, the longest one about 5 × IIIbd. Apical segm. of rostrum 1.1–1.5 × 2sht., with 3–5, usually 4, accessory hairs. First tarsal segm. typically with 5 hairs. Cauda distinctly constricted (Fig. 262). 1.3–1.8 mm, the smallest individuals occur in summer.

Fundatrix. Black with pale median stripe. Antenna shorter than 0.5 × body; processus terminalis 1.3–1.5 × VIa. 1.9–2.2 mm.

Alate viviparous female. Black. Abdomen with sclerotic dorsal cross bars and marginal sclerites. Secondary rhinaria on ant. segm. III: 7–12, IV: (0–)1–4, V: 0.

Oviparous female. Light brown with head and dorsal abdominal markings, darker. Antennae and legs dark. Siphunculi pale at base. Antenna shorter than 0.5 × body; processus terminalis 2–2.3 × VIa. Cauda very slightly constricted. Hind tibiae swollen, with numerous scent plaques. Body length as in fundatrix.

Apterous male. Black. Abdomen with cross bars and marginal sclerites. Secondary rhinaria on ant. segm. III: 5–17, IV: 5–12, V: 1–7.

Distribution. In Denmark uncommon, known from NWJ, NEJ and NEZ; in Sweden widespread in the southern part of the country, north to Med.; not in Norway; in Finland known from Ab, Ta, Sb and Kb. – Widespread in Europe and Asia, south to Spain, Hungary, and Caucasus, east to W Siberia and C Asia; rare in Great Britain; not rare in N Germany; recorded from the Baltic region of Poland.

Biology. The species lives on the undersides of leaves of *Salix cinerea, caprea,* and *aurita.* It is sometimes visited by ants.

134. ***Chaitophorus salijaponicus*** Essig & Kuwana, 1918

Chaitophorus salijaponicus Essig & Kuwana, 1918: 85. – Survey: 142.

The species is subdivided into two subspecies, viz. *s. salijaponicus* from the Far East and *s. niger* from Europe and N Asia.

Chaitophorus salijaponicus niger Mordvilko, 1929

Figs. 247, 263, 273, 278, 279.

Chaitophorus niger Mordvilko, 1929: 29. – Survey: 142.

Apterous viviparous female. Black; with a pale ring around base of siphunculus (Fig. 278). Abd. tergites (I–)II–VI fused. Dorsal cuticle with wavy rows of denticles (much more distinct than in *salicti*) that tend to form reticulate patterns on thorax and anterior part of abdomen (Fig. 279). Dorsal hairs long, about 6–8 × IIIbd., pointed. Antenna a little longer than 0.5 × body; processus terminalis 2.1–3.0 × VIa, a little longer than segm. III; segm. III with 7–11 hairs, 4–5 × IIIbd. Apical segm. of rostrum 0.8–1.1 × 2sht., with 2(–4) accessory hairs. First tarsal segm. with 5(–6) hairs. Cauda distinctly constricted (Fig. 263). 1.2–2.2 mm.

Fundatrix. Similar to the apterae of later generations, but processus terminalis only 1.7–2.1 × VIa.

Alate viviparous female. Black. Abdomen with marginal sclerites and more or less fused dorsal cross bars (Fig. 247). Secondary rhinaria on ant. segm. III: 7–10, not arranged in a row, IV: 0–2, V: 0.

Oviparous female. Head and anterior part of thorax dark pigmented. Abdomen membranous, light brown, with marginal sclerites on segm. VI, and dorsal cross bars on segm. VII–VIII, dark. Cauda more or less distinctly constricted. Hind tibiae strongly swollen, with numerous scent plaques. Rather large, 2.0–2.3 mm.

Male. Alate or apterous. Black. Antenna almost as long as body; secondary rhinaria on segm. III: 6–31, IV: 9–22, V: 5–15. First tarsal segm. with 6 or 7 hairs.

Distribution. In Sweden found in Sk. and Upl.; in Norway in Bø and HOy; not in Denmark and Finland. – Widespread in Europe and Asia, south to Spain, Turkey,

Ukraine, and Caucasus, east to C Asia, Mongolia, and the Ussuri region; local and uncommon in Great Britain; known from the Netherlands, N Germany, Poland, and NW & W Russia.

Biology. The species lives in small colonies on leaves of various *Salix* spp., e.g. *alba, fragilis, amygdalina, babylonica, purpurea,* and *laurina.* It is sometimes visited by ants in the spring.

135. ***Chaitophorus tremulae*** Koch, 1854
Plate 4: 9. Fig. 280.

Chaitophorus tremulae Koch, 1854: 8. – Survey: 143.

Apterous viviparous female. Black, very often with pale longitudinal median stripe. Antennae (except basal parts of flagellum) and siphunculi black. Abd. tergites I–VI fused. Dorsal cuticle with numerous small, triangular, denticular spinules or flat nodules (Fig. 280). Dorsal hairs pointed, 0.04–0.10 mm long. Antenna longer than 0.5 × body; processus terminalis 2.1–2.8 × VIa, about as long as segm. III; segm. III with 8–13 hairs, the longest one 1.8–2.8 × diameter of the segment. Apical segm. of rostrum as long as, or shorter than, 2sht., with (3–)4(–6) accessory hairs. First tarsal segm. with 7 hairs. Cauda distinctly constricted. 1.3–2.5 mm.

Fundatrix. Not blackish sclerotic. Abd. tergite I free. Alate viviparous female. Head, thorax, antennae, legs, and siphunculi, black. Abdomen dark green, with black dorsal cross bars, and equally dark marginal sclerites. Cross bars on tergites III–VI broad, solid, nearly fused into a solid mid-dorsal patch. Antenna 0.65–0.75 × body; 10–23

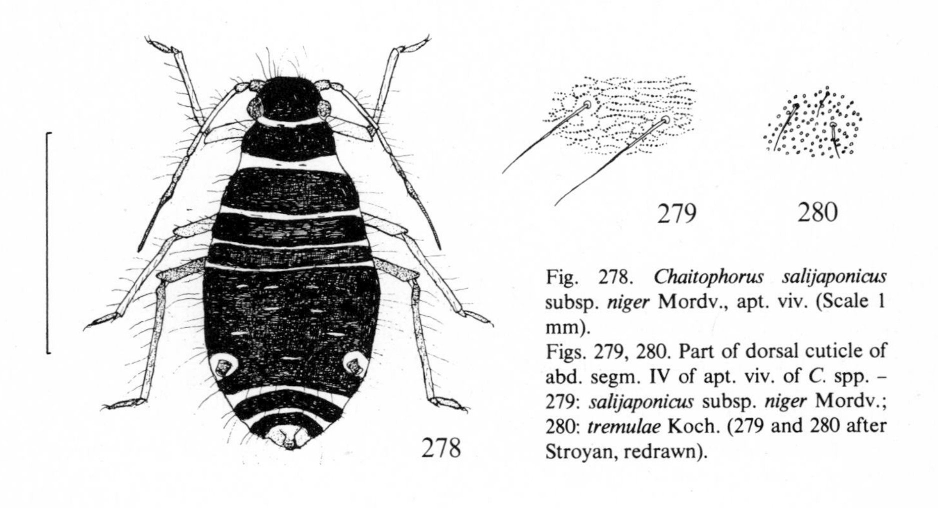

Fig. 278. *Chaitophorus salijaponicus* subsp. *niger* Mordv., apt. viv. (Scale 1 mm).
Figs. 279, 280. Part of dorsal cuticle of abd. segm. IV of apt. viv. of *C.* spp. – 279: *salijaponicus* subsp. *niger* Mordv.; 280: *tremulae* Koch. (279 and 280 after Stroyan, redrawn).

secondary rhinaria on segm. III, irregularly arranged in a double, or partly triple, row along one side, IV: 0–4, V: 0; segm. III with 19–24 hairs on both antennae together.

Oviparous female. Dirty yellowish green, with darker head and pronotum. Siphunculi pale at base. Body elongate, spindle-shaped. Hind tibia swollen, with numerous scent plaques. Cauda wart-shaped, dark. Rather large, 2.0–2.6 mm.

Alate male. Black. Abdomen with marginal sclerites and dorsal cross bars. Antenna about 0.75 × body; processus terminalis 2.6–3.3 × VIa; secondary rhinaria on segm. III: 35–51, IV: 18–29, V: 9–24.

Distribution. In Denmark rather common in Jutland: EJ, WJ, NWJ, and NEJ; in Sweden common and widespread from Sk. in the south to P.Lpm. in the north; in Norway common, north to Nsy; in Finland known from Ab, N, Ta, and Ok. – All over Europe, including the British Isles, the Netherlands, N Germany, Poland, and NW & N Russia, south to Spain and Hungary, east to Ural. Asia: Turkey, Caucasus, Kazakhstan, W Siberia, Mongolia.

Biology. Small colonies are found on the undersides of leaves of *Populus tremula,* often between leaves spun together by lepidopterous larvae. When disturbed, the aphids escape rather quickly like bugs. They are not attended by ants.

136. ***Chaitophorus truncatus*** (Hausmann, 1802)
Fig. 274.

Aphis truncata Hausmann, 1802: 443. – Survey: 143.

Apterous viviparous female. Light green, sometimes with three dark green longitudinal dorsal stripes, or blackish; the dark individuals mostly occur in early summer and in autumn. Abd. tergites II–VI fused. Dorsal cuticle with flat nodules; dark areas weakly and sparsely nodulose, pale areas hardly sculptured. Dorsal hairs long, about 4–6 × IIIbd., in some individuals (especially from spring and summer) with blunt or expanded, sometimes furcate, apices; other individuals (especially from late summer and autumn) with pointed hairs only. Antenna longer than 0.5 × body; processus terminalis 2.2–3.2 × VIa, usually a little longer than segm. III, rarely a little shorter; segm. III with 6–11 hairs, 2–4 × IIIbd. Apical segm. of rostrum 0.6–0.9 × 2sht., with 2, exceptionally 3, accessory hairs. First tarsal segm. with 7 hairs. Cauda distinctly knobbed 1.4–2.2 mm.

Alate viviparous female. Head, thorax, antennae, and siphunculi, dark. Abdomen light green, with dark green cross bars strongly broken up and not forming a mid-dorsal patch. Dorsal hairs pointed. Antenna about 0.75 × body; secondary rhinaria on segm. III: 6–22 in a single or an irregularly double row along one side, IV: 0–3, V: 0–2; segm. III with 16–19 hairs on both antennae together.

Oviparous female. Head and thorax lightly pigmented. Abd. tergite VIII with trace of dark pigmentation. Body spindle-shaped. Dorsal hairs pointed. Antenna about 0.5 × body. Hind tibia dark, with numerous scent plaques. Cauda sometimes without constriction. Rather large, 2.2–2.7 mm.

Apterous male. Black. Abdomen with marginal sclerites and dorsal cross bars. Antenna a little shorter than body; secondary rhinaria on segm. III: 8–20, IV: 4–13, V: 3–8. Rather small, 1.3–1.6 mm.

Distribution. In Denmark known only from NEZ (Tåstrup, trap); not in Sweden; in Norway found in On (Vaset Seter near Fagernes, Heie, unpublished); in Finland widespread from N in the south to ObN in the north. – The British Isles, the Netherlands, Germany, Poland, Russia, Czechoslovakia, Hungary; W Kazakhstan. In Great Britain widespread and fairly common; in N Germany not rare.

Biology. The hosts are various *Salix* spp., e.g. *purpurea, amygdalina, alba, triandra, babylonica,* and *phylicifolia.* Small colonies are found on the leaves. They are not visited by ants.

137. ***Chaitophorus vitellinae*** (Schrank, 1801)
Figs. 248, 281–284.

Aphis vitellinae Schrank, 1801: 103. – Survey: 143.

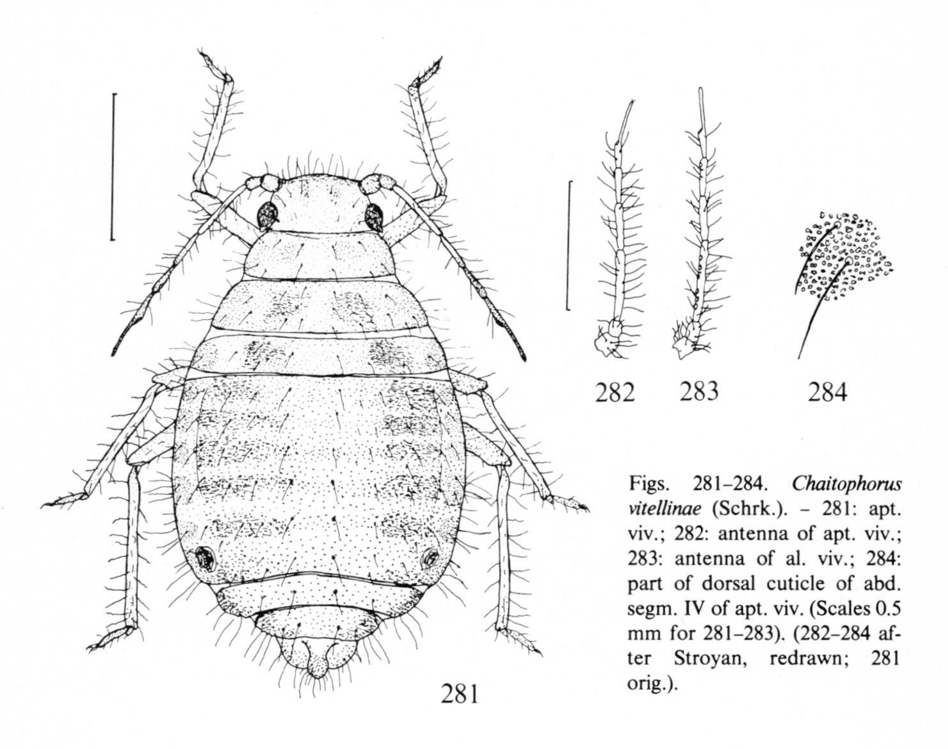

Figs. 281–284. *Chaitophorus vitellinae* (Schrk.). – 281: apt. viv.; 282: antenna of apt. viv.; 283: antenna of al. viv.; 284: part of dorsal cuticle of abd. segm. IV of apt. viv. (Scales 0.5 mm for 281–283). (282–284 after Stroyan, redrawn; 281 orig.).

Apterous viviparous female. Yellowish green with two darker green longitudinal dorsal stripes. Dorsum rarely with dark pigmentation. Antennae dark at bases and at apices. Abd. tergites (I–)II–VI fused . Dorsal cuticle with dense, coarse, irregular nodular sculpture (Fig. 284). Dorsal hairs erect, pointed, 3–5 × IIIbd. Antenna longer than 0.5 × body; processus terminalis 1.4–2.0 × VIa, shorter than segm. III; segm. III with 10–23 hairs, the longest one about 3 × IIIbd. (Fig. 282); the longer antennal hairs rather evenly distributed round circumference of segments (as in *populeti*). Apical segm. of rostrum shorter than hind tarsus, 1.1–1.2 × 2sht., with 4–9 accessory hairs. First tarsal segm. with (6–)7 hairs. Cauda usually only slightly constricted. 1.5–1.9 mm.

Fundatrix. Grass green. Antenna shorter than 0.5 × body; some segments often fused; processus terminalis only a little longer than basal part of ultimate segm.

Alate viviparous female. Head and thorax black. Abdomen green, with marginal sclerites and rather narrow, evenly spaced, dorsal cross bars (Fig. 248). Siphunculi pale. Secondary rhinaria on ant. segm. III: 3–10, IV: 0–2, V: 0 (Fig. 283).

Oviparous female. Yellowish green, with head, pronotum, apices of ant. segments, and siphunculi darker. Antenna shorter than 0.5 × body. Hind tibia somewhat swollen, with numerous scent plaques. Rather large, 2.2–2.5 mm.

Apterous male. Dark brown, nearly black. Antennae dark. Abdomen with marginal sclerites and dorsal cross bars. Antenna about 0.75 × body; processus terminalis 1.1–1.4 × VIa; segm. III with 9–12 rhinaria, IV with 5–11, V with 1–5 secondary rhinaria.

Distribution. In Denmark known only from NEZ (Tåstrup, in trap); in Sweden found in Sk.; not in Norway or Finland. – Great Britain (rare), the Netherlands, Germany, Poland, Russia, Switzerland, Austria, Czechoslovakia, Hungary, Italy, Yugoslavia; Caucasus, Kazakhstan, W Siberia.

Biology. The host plants are *Salix alba, fragilis, viminalis,* and *amygdalina.* The aphids live in colonies on young shoots, especially at bases of leaf stems, and on twigs, and are visited by ants (especially *Lasius fuliginosus*).

TRIBE SIPHINI

Body oval, elongate, sometimes slender med almost parallel-sided, more or less long-haired. Antenna very short, usually 5-segmented, rarely 4-segmented, without secondary rhinaria in apterous females; alate females with rather few small secondary rhinaria on ant. segm. III. Rostrum short. Siphunculi low, stump-shaped, conical, or pore-shaped, never with reticulate sculpture, placed on margins of abd. segm. V or VI. Scent plaques on hind tibiae of oviparous females often fused two by two. Males apterous (if known).

The aphids live on stems, leaves, and inflorescences of herbaceous monocotyledones, viz. Gramineae and Cyperaceae (rarely on Juncaceae). Most species are holocyclic.

Key to genera of Siphini

Apterous and alate viviparous females

1 Siphunculi placed on abd. segm. VI (Fig. 288). 2
– Siphunculi placed on abd. segm. V (Fig. 302). 3
2 (1) Compound eyes placed on lateral peduncle-like outgrowths of the head (Fig. 286). Body broadly pear-shaped. Cauda constricted. Siphunculi truncate conical or stump-shaped. *Caricosipha* Börner (p. 142)
– Compound eyes not placed on lateral outgrowths of the head (Fig. 287). Body elongate, slender. Cauda not constricted. Siphunculi only slightly raised pores. .... *Laingia* Theobald (p. 144)
3 (1) Apical segm. of rostrum stylet-shaped, longer than ant. segm. III (Figs. 320, 321). .... *Chaetosiphella* Hille Ris Lambers (p. 155)
– Apical segm. of rostrum not stylet-shaped, shorter than ant. segm. III. 4
4 (3) Body elongate, almost parallel-sided; body length (when mounted on slides) 2.5 × greatest width, or more. Siphunculi small, pore-shaped. .... *Atheroides* Haliday (p. 145)
– Body less elongate, oval. Siphunculi elevated on low rim-like truncate cones, or compressed vasiform. .... *Sipha* Passerini (p. 149)

Genus *Caricosipha* Börner, 1939

Caricosipha Börner, 1939: 77.
Type-species: *Caricosipha paniculatae* Börner, 1939.
Survey: 126.
Only one species.

138. ***Caricosipha paniculatae*** Börner, 1939
Plate 4: 15. Figs. 285, 286.

Caricosipha paniculatae Börner, 1939: 77. – Survey: 126.

Apterous viviparous female. Yellow or reddish, with dark brown sclerotic areas, or black over entire dorsum and most of venter. Antennae and legs brownish yellow. Siphunculi palish. Body flattened, broadly pear-shaped, broadest about level of abd. segm. IV. Dorsum sclerotic, often with paler longitudinal median stripe, with very long, black, pointed hairs and fine spinules. Borderlines between head and pronotum, and between abd. tergites I–VII, not distinctly visible. Frons convex; longest hair on frons longer than ant. segm. IV, sometimes longer than segm. III. Eyes very prominent, placed on constricted peduncular lateral extensions of head (Fig. 286). Ocular tubercles distinct. Antenna about 0.5 × body, or a little longer, 5-segmented; processus terminalis about 2.5 × Va; segm. III with 3–5 long erect hairs on inner face of segm., segm. IV with one long hair; the longest antennal hairs about 5 × IIIbd; the basal hair on segm. III often thinner and shorter than the others; additional short normal hairs present on segm. IV and V. Apical segm. of rostrum short, blunt, with 2 accessory hairs. First tarsal segm. with 5 hairs. Empodial hairs spatulate. Siphunculi short, stump-shaped, smooth, with flange, placed on abd. segm. VI. Cauda knobbed. Anal plate emarginate. 1.5–2.4 mm.

Alate viviparous female. Abdomen with marginal sclerites and dorsal cross bars. Body more slender, with shorter and thinner hairs than in the apterous viviparous female. Ant. segm. III with 9–22 rhinaria. Wings narrow; media of fore wing with one fork.

Oviparous female. Similar to the apterous viviparous female. Hind tibia not swollen or only slightly so, with 4–40 8-shaped scent plaques.

Apterous male. Similar to the viviparous females, but slenderer, with relatively longer hairs. Antenna about 0.65 × body; processus terminalis almost 3 × Va; secondary rhinaria on segm. III: 22–35, IV: 6–11.

Distribution. In Denmark found in EJ, WJ, NWJ, and NEZ; in Sweden in Sk., Öl., Gtl., Ög., Nrk., Sdm., Upl., and Lu.Lpm.; in Norway in AK; not in Finland. – Great Britain, the Netherlands, Germany, Poland, Russia, France, Spain, Czechoslovakia, Hungary.

Biology. The aphids live on leaves of *Carex* spp. in meadows, at lake shores, and in marshy places. The main host is *C. paniculata*, but also frequented are *C. paniculata* × *remota, vulpina, disticha, flava, atrata,* and *C.* sp. They are brisk insects which are able to run rather fast. They are not visited by ants.

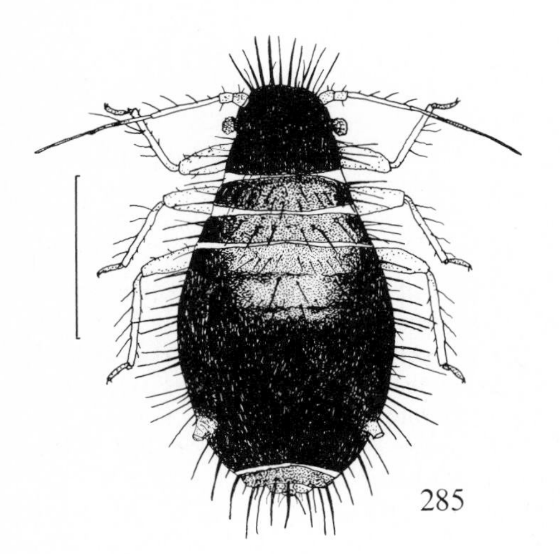

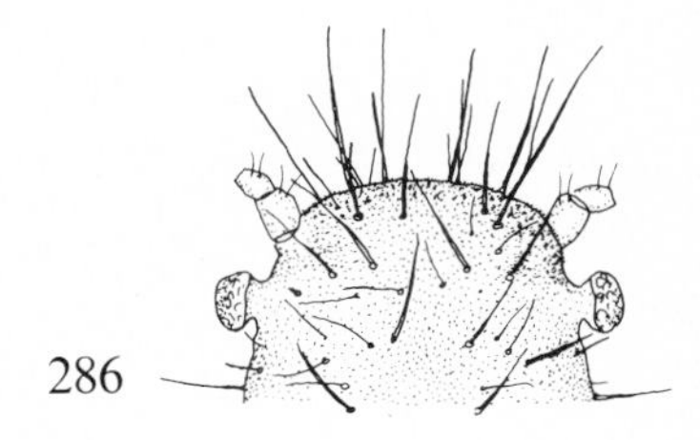

Figs. 285, 286. *Caricosipha paniculatae* Börner, apt. viv. – 285: habitus; 286: head and anterior part of pronotum. (Scale 1 mm for 285). (286 after Richards, redrawn).

Genus *Laingia* Theobald, 1922

Laingia Theobald, 1922: 429.
Type-species: *Laingia psammae* Theobald, 1922.
Survey: 240.

Only one species.

139. ***Laingia psammae*** Theobald, 1922
Plate 4: 12. Figs. 287, 288.

Laingia psammae Theobald, 1922: 429. – Survey: 240.

Apterous viviparous female. Dirty straw-yellowish to greyish green. Body elongate, more than twice as long as broad. Dorsum sclerotic with distinct membranous intersegmental lines, densely covered with wavy rows of fine spinules, laterally forming a reticulate pattern. Dorsal hairs short, spiny, the longest ones about 2 × IIIbd.; frons and posterior part of abdomen with longer hairs, 6 × IIIbd., or longer. Frons convex. Ocular tubercles well developed. Antenna about 0.25 × body; processus terminalis 1.2–2.1 × Va; Va as long as, or a little longer than, segm. IV; segm. III with 2–3 long hairs, IV with 1; longest hair on segm. III 1–1.75 ×IIIbd. Apical segm. of rostrum about 0.75 × 2sht. Legs short. First tarsal segm. with 5 hairs. Empodial hairs simple. Siphunculus low, almost pore-shaped, placed on abd. segm. VI; diameter of aperture a little longer than width of hind tibia in the middle. Cauda and anal plate broadly rounded. 1.8–2.8 mm.

Alate viviparous female. Brown or brownish green. Abdomen with marginal sclerites and dark dorsal cross bars. Ant. segm. III as long as, or a little longer than, segm. V, with 3–7 rhinaria in a row; the longest hairs about 2 × IIIbd. Wings narrow; media of fore wing with one or two forks. Cauda almost squarish.

Distribution. In Denmark found in WJ, NWJ, and NEJ, common in the dunes along the west coast of Jutland; in Sweden in Sk., Hall., Gtl., Dlsl., Nrk., Upl., Vstm., and Vrm.; not in Norway; in Finland in N and Sa. – In Europe and Asia, south to Spain, east to W Siberiaand C Asia; in Great Britain common round the coast, less common on inland localities; common in N Germany; known from the Baltic region of Poland and NW Russia.

Biology. The aphids live between the flowers, or the fruits, in the inflorescences of the grasses *Ammophila arenaria* and *Calamagrostis epigeios;* in Sweden also collected on *Elymus, Calamagrostis arundinacea,* and *Deschampsia caespitosa.* It is common on *Ammophila* along the North Sea coast of Jutland, less common on *Calamagrostis epigeios* on inland localities, while most Swedish records are from *Calamagrostis.* The inflorescences of the infested grasses often become sticky with honeydew and more or less blackish. The species is not visited by ants. The sexuales have not been described.

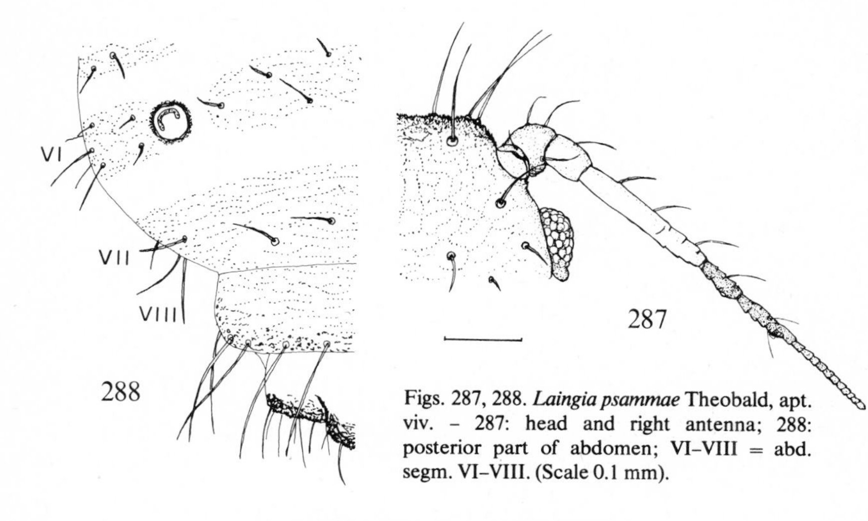

Figs. 287, 288. *Laingia psammae* Theobald, apt. viv. – 287: head and right antenna; 288: posterior part of abdomen; VI–VIII = abd. segm. VI–VIII. (Scale 0.1 mm).

Genus *Atheroides* Haliday, 1839

Atheroides Haliday, 1839: 189.
Type-species: *Atheroides serrulatus* Haliday, 1839.
Survey: 98.

Body elongate, often nearly parallel-sided, more than twice as long as broad. Dorsum sclerotic, yellow, brown or black; head and pronotum without membranous borderline; also abd. tergites II–VII usually fused. Frons convex or straight. Antenna 4- or 5-segmented, shorter than 0.3 × body. Rostrum short; apical segm. terminating into a beak-like structure. Legs Very thin. Empodial hairs spatulate (except in *brevicornis*). Siphunculi pore-shaped, placed on abd. segm. V. Cauda and anal plate broadly rounded. Cauda usually covered by abd. tergite VIII, not visible from above.

Five species in the world, four species in Scandinavia. They live on grasses and are not visited by ants.

Key to species of *Atheroides*

Apterous viviparous females

1 Some of the dorsal hairs long and pointed, the longest ones as long as basal part of ultimate ant. segm., or longer; others short, blunt or spine-like, sometimes furcate. On *Deschampsia caespitosa*. 2
– Dorsal hairs apparently absent or very short, shorter than 0.33 × basal part of ultimate ant. segm. Longer hairs present only on frons and on posterior part of abdomen. On various grasses. 3

2 (1) Body parallel-sided. Dorsum yellowish brown. Antenna about 0.22 × body. .... 141. *doncasteri* Ossiannilsson

– Body broadest across anterior part of abdomen. Dorsum black. Antenna about 0.26–0.29 × body. .... 142. *hirtellus* Haliday

3 (1) Antenna longer than 0.20 × body, 5-segmented. Processus terminalis about 0.75–1.0 × basal part of ultimate ant. segm. Each abd. tergite with 2 pairs of spinal hairs and 1 pair of pleural hairs. Empodial hairs spatulate. Colour in life yellow or brownish. .... 143. *serrulatus* Haliday

– Antennae shorter than 0.20 × body, usually 4-segmented, seldom 5-segmented. Processus terminalis almost 0.5 × basal part of ultimate ant. segm. Each abd. tergite with several irregularly scattered dorsal hairs. Empodial hairs simple. Colour in life brownish or black. .... 140. *brevicornis* Laing

140. ***Atheroides brevicornis*** Laing, 1920
Fig. 289.

Atheroides brevicornis Laing, 1920: 41. – Survey: 98.

Apterous viviparous female. Brownish to black. Antennae and legs dark brown. Margins of body slightly convex (Fig. 289). Dorsal cuticle sclerotic, coarsely wrinkled, rugose; abdomen without membranous intersegmental lines on tergites II–VII. Dorsal hairs very short, irregularly scattered in 2–3 transverse rows across tergites, flabellate or club-shaped, with blunt or ragged apices; abd. tergite VIII with additional long, thick, pointed marginal hairs. Ocular tubercles inconspicuous. Antenna 0.14–0.17 × body, usually 4-segmented, rarely 5-segmented; processus terminalis almost as long as 0.5 × IVa (Va); antennal hairs shorter than 0.5 × IIIbd. Apical segm. of rostrum as long as 2sht. First tarsal segm. with 2 or 3 hairs. Empodial hairs simple. About 2 mm.

Alate viviparous female. Abdomen with marginal sclerites and more or less fused spinal sclerites, sometimes also with pleural sclerites; tergites VII and VIII entirely sclerotic. Dorsal hairs thinner than in apterae. Antenna 4- or 5-segmented; segm. III with 2–4 rhinaria. Venation of wings variable; media of fore wing with one or two forks; cubitus more or less branched.

Distribution. In Sweden known from Boh. (Stenungsund); in Norway trapped in HOy; not in Denmark or Finland. – Great Britain, the Netherlands, Germany, Hungary, Crimea.

Biology. The species lives on the uppersides of leaves of *Festuca* and *Puccinellia* in coastal saltings. Ossiannilsson (1959) found it in Sweden on *Puccinellia distans*.

141. ***Atheroides doncasteri*** Ossiannilsson, 1955
Figs. 290–292.

Atheroides doncasteri Ossiannilsson, 1955: 127. – Survey: 98.

Apterous viviparous female. Dirty yellowish to brownish. Body almost parallel-sided (Fig. 290). Dorsum sclerotic; abdomen without membranous intersegmental lines on tergites II–VII; with many long hairs, the longest one longer than ultimate ant. segm., and still more numerous shorter hairs having blunt or furcate apices (Fig. 291). Antenna about 0.22 × body, 5-segmented; processus terminalis about as long as Va; longest antennal hair a little shorter than ant. segm. II. 1.7–2.1 mm.

Alate viviparous female. Darker than the apterous viviparous female. Abdomen with marginal sclerites and dorsal cross bars. Ant. segm. III with 2–4 rhinaria on distal $^{2}/_{3}$. Wing veins dark-bordered.

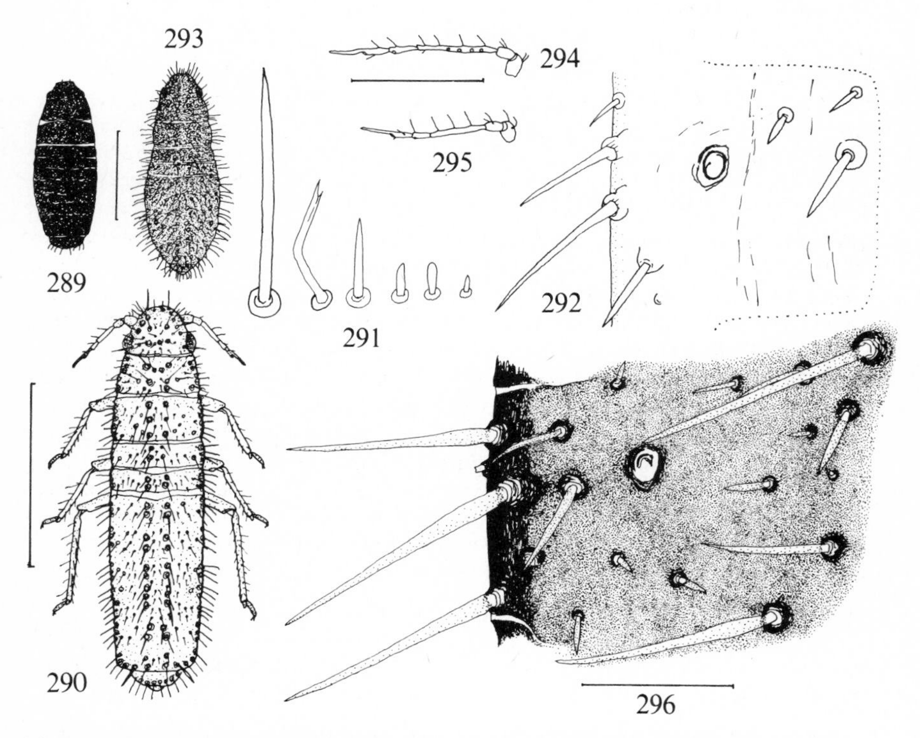

Fig. 289. *Atheroides brevicornis* Laing, body of apt. viv. (Scale 1 mm). (After Stroyan, redrawn).
Figs. 290–292. *A. doncasteri* Ossiann. apt. viv. (paratype). – 290: habitus; 291: types of dorsal hairs; 292: left margin of abd. tergite V with siphuncular pore. (Scale 1 mm for 290).
Figs. 293–296. *A. hirtellus* Haliday. – 293: body of apt. viv.; 294: antenna of al. viv.; 295: antenna of apt. viv.; 296: left margin of abd. tergite V of apt. viv. with siphuncular pore. (Scales 1 mm for 293, 0.5 mm for 294 and 295, 0.1 mm for 296). (293–295 after Stroyan, redrawn; 296 orig.).

Distribution. In Sweden found in Nrk. (Örebro), Sdm. (Mariefred), and Upl. (Uppsala-Näs and Vaksala); not in Denmark, Norway, or Finland. – The Netherlands, Czechoslovakia, Hungary, W Siberia.

Biology. The host is *Deschampsia caespitosa.*

142. ***Atheroides hirtellus*** Haliday, 1839
Figs. 293–296.

Atheroides hirtellus Haliday, 1839: 189.
Atheroides niger Ossiannilsson, 1954b: 117.
Survey: 98.

Apterous viviparous female. Black with underside dirty yellowish. Antennae and legs yellowish. Body lanceolate, broadest about level of abd. segm. II (Fig. 293). Dorsal cuticle sclerotic, wrinkled, not rugose; abdomen without membranous intersegmental lines on tergites II–VII. Dorsum with many long hairs (Fig. 296), on abd. tergite III 0.11–0.18 mm long, equal to or longer than Va, placed on very small, pale spots; also shorter spiny hairs. Antenna 0.26–0.29 × body, 5-segmented (Fig. 295); processus terminalis 1.1–1.2 × Va; longest hair on segm. III 0.05–0.07 mm, 0.5 × Va or longer, about as long as segm. II. First tarsal segm. typically with 5 hairs. Hind margin of abd. tergite VII more or less straight, not covering cauda as in the other species of *Atheroides.* 2.0–2.8 mm.

Alate viviparous female. Abdomen with marginal, and more or less fused, dorsal sclerites. Ant. segm. III with 1–4 rhinaria (Fig. 294).

Oviparous female. Paler than apterous viviparous female. Hind tibia slightly thickened, with many 8-shaped scent plaques.

Distribution. In Sweden found in Sm. (Alvesta), Nrk. (Örebro), and Sdm. (Åby); in Finland known from N; not in Denmark or Norway. – Great Britain, Germany (not N Germany), Poland (not the Baltic region), and France (Languedoc).

Biology. The host is *Deschampsia caespitosa.*

143. ***Atheroides serrulatus*** Haliday, 1839
Plate 4: 13. Figs. 297–299.

Atheroides serrulatus Haliday, 1839: 189. – Survey: 99.

Apterous viviparous female. Yellow or yellowish brown. Body almost parallel-sided. Dorsal cuticle strongly wrinkled, rugose (Fig. 298). Dorsum with very few short hairs, less than 0.03 mm long, rod-shaped, with blunt to ragged apices (Fig. 299). Hairs on frons and margins of posterior body segments longer; abd. tergite VIII with 14–20 very long, thick, pointed hairs (Fig. 297). Antenna hardly as long as 0.25 × body, 5-segmented; processus terminalis 0.8–1.2 × Va; antennal hairs about as long as IIIbd. Apical segm. of rostrum shorter than 2sht. First tarsal segm. with 5, rarely 4, hairs. 1.9–2.4 mm.

Alate viviparous female. Abdomen with dark marginal sclerites and broad dorsal cross bars; posterior tergites entirely sclerotic. Processus terminalis about 1.4 × Va. Secondary rhinaria on ant. segm. III: 4–5 along the middle part of segm., IV: 0–1.

Oviparous female. Similar to the apterous viviparous female, but hind tibia slightly thickened, with about 50 8-shaped scent plaques placed over almost entire surface.

Distribution. Common and widespread in Denmark; in Sweden common, known from most districts, from Sk. in the south to Lu.Lpm. in the north; in Norway recorded from AK, Ry, HOy, and HOi; in Finland from N and Oa. – Europe and Asia, south to Spain, Yugoslavia, and Turkey, east to W Siberia; common and widespread in Great Britain, N Germany, Poland, and NW & W Russia. Richards (1972) recorded it from Canada (Quebec).

Biology. The species lives on the leaves of various grasses, e.g. *Poa, Agrostis, Festuca, Deschampsia, Alopecurus,* and *Nardus.*

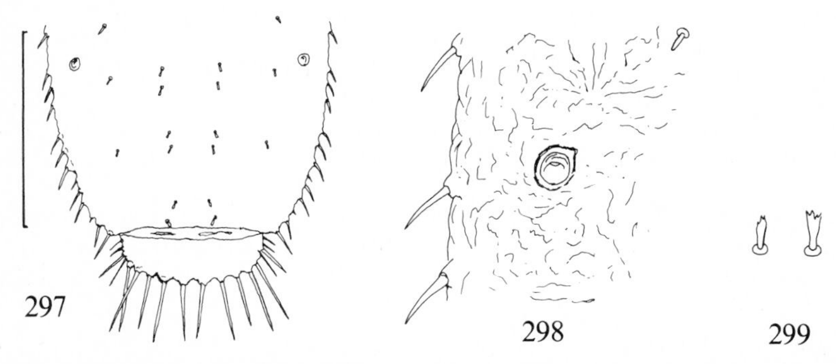

Figs. 297–299. *Atheroides serrulatus* Haliday, apt. viv. – 297: posterior part of abdomen, dorsal view; 298: left margin of abd. tergite V with siphuncular pore; 299: shape of mid-dorsal hairs. (Scale 0.5 mm for 297).

Genus *Sipha* Passerini, 1860 s. lat.

Sipha Passerini, 1860: 29.

Type-species: *Aphis glyceriae* Kaltenbach, 1843.

Survey: 397.

Body oval to elongate oval, rather flat. Dorsum of apterae sclerotic, more or less dark pigmented, usually with membranous intersegmental lines only between head, thoracic tergites, and abd. tergites I–II and VII–VIII. Dorsal hairs pointed or blunt. Antenna 0.5 × body or shorter, 5-segmented. Rostrum short, reaching about to middle coxae; apical segm. rather short, subtriangular, with rather acute apex (Figs. 313–316). Empodial hairs apparently simple, a little flattened in one plane. Siphunculi short, stump-shaped, placed on abd. segm. V.

Vith 12 species in the world, 5 species in Scandinavia. They live on grasses (*glyceriae* occasionally also on other monocotyledones), with or without attention by ants. The genus is holarctic and subdivided into two subgenera.

Key to subgenera of *Sipha*

Apterous and alate viviparous females

1 Cauda constricted, more or less knobbed (Fig. 311). . *Sipha* Passerini s. str. (p. 150)
– Cauda not constricted, broadly rounded (Fig. 312). .... *Rungsia* Mimeur (p. 152)

Subgenus *Sipha* Passerini, 1860 s. str.

Dorsal cuticle of apterae (of European species) with numerous blunt nodules, or pointed denticles, or spinules, often arranged in rows forming an indistinct reticulate pattern (Figs. 303, 304). Distance between eye and antennal base about as long as eye (Fig. 300). First tarsal segm. with 3 or 4 hairs. Cauda constricted, more or less knobbed (Fig. 311).

Four species in the world, two species in Scandinavia.

Key to species of *Sipha* s. str.

Apterous viviparous females

1 Longest hair on abd. tergite III longer than 0.045 mm. Longest hair on ant. segm. III usually longer than IIIbd. (Fig. 308).
144. *glyceriae* (Kaltenbach)
– Longest hair on abd. tergite III 0.04 mm, or shorter. Longest hair on abd. tergite III 0.04 mm, or shorter. Longest hair on ant. segm. III as long as IIIbd., or shorter. (Fig. 309). .... 145. *littoralis* (Walker)

144. ***Sipha (Sipha) glyceriae*** (Kaltenbach, 1843)
Plate 4: 10. Figs. 300, 303, 305, 308, 311, 313.

Aphis glyceriae Kaltenbach, 1843: 113. – Survey: 397.

Apterous viviparous female. Dull green or yellowish, without dark spots. Dorsal cuticle with small, pointed denticles between the long, pointed, thorn-like hairs (Figs. 303, 305). Antenna about 0.33 × body; processus terminalis shorter or longer than Va, longer than 0.5 × segm. III; segm. III with 1–4 hairs, 0.6–1.8 × IIIbd. (Fig. 308). 1.7–2.5 mm.

Alate viviparous female. Head and thorax dark. Abdomen with dark marginal and dorsal sclerites. Antenna about 0.4 × body; segm. III with about 2 rhinaria in the middle part. Wings narrow.

Oviparous female. Hind tibiae swollen, with rather few inconspicuous scent plaques on basal ¾. Cauda only slightly constricted. Otherwise much like the apterous viviparous female.

Apterous male. Greenish brown, reddish brown, or blackish, with paler segmental borders. Secondary rhinaria on ant. segm. III: about 30–32, IV: about 8–9. Rather small, 0.7–1.5 mm.

Distribution. Common and widespread in Denmark; common in Sweden north to Hls.; in Norway known from HEs, Ry, HOy, Hoi, and SFi; in Finland from Ta, Sa, Oa, and Ok. – Europe and Asia, south to Spain, Turkey, and Transcaucasia, east to W

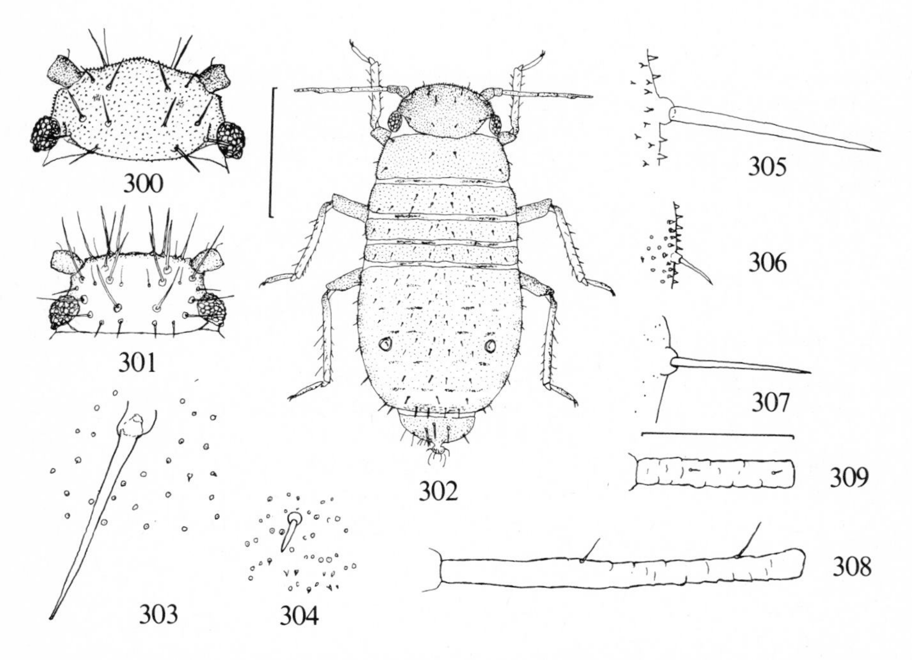

Figs. 300, 301. Heads of *Sipha glyceriae* (Kalt.) (300) and *S. (Rungsia) elegans* (d. Guerc.) (301). (Scale 0.3 mm).
Fig. 302. *S. littoralis* (Wlk.), apt. viv. (Scale 0.5 mm).
Figs. 303, 304. Part of dorsal cuticle of *Sipha* spp. – 303: *glyceriae;* 304: *littoralis.*
Figs. 305–307. Part of right margin of abd. tergite III of *Sipha* spp. – 305: *glyceriae;* 306: *littoralis;* 307: *maydis.* (Scale 0.1 mm).
Figs. 308, 309. Ant. segm. III of *Sipha* spp., apt. viv. – 308: *glyceriae;* 309: *littoralis.* (Scale 0.1 mm).

Siberia and C Asia; very common in Great Britain; not rare in N Germany; known from the Baltic region of Poland and NW & W Russia. N America: Known from the eastern part of the USA and Canada, and also from British Columbia.

Biology. The aphids live on the leaves of various grasses, e.g. *Glyceria fluitans* and species of *Poa, Phleum, Agrostis, Festuca,* and *Agropyrum,* especially where growing in wet habitats. Rarely also found on some Cyperaceae or Juncaceae in such habitats. It can occasionally be found on barley (*Hordeum*) and other cereals. It is sometimes visited by ants.

145. ***Sipha (Sipha) littoralis*** (Walker, 1848)
Figs. 302, 304, 306, 309.

Aphis littoralis Walker, 1848b: 44. – Survey: 397.

Apterous viviparous female (Fig. 302). Green, dark green, or bluish green, with somewhat brownish head and thorax. Antennae dark. Legs brown with somewhat paler tibiae. Abdomen with brownish spots. Siphunculi brownish. Cauda pale. Dorsal cuticle on head and thorax, and around bases of the rather short abdominal hairs(Figs. 304, 306), with small, irregularly shaped, nodular denticles, with blunt or roundishapices. Antenna about 0.25 × body; processus terminalis a little shorter than Va; segm. III with 1–2 hairs distally, 1.0 × IIIbd. or shorter (Fig. 309). 1.4–2.2 mm.

Apterous male. Brown. Antenna about 0.5 × body.

Distribution. In Denmark found in the marsh at the North Sea in SJ (Højer); in Sweden known from Nb.; not in Norway or Finland. – NW Europe: Great Britain, the Netherlands, and NW Germany. The Swedish record seems to be the only one from an area outside the North Sea area.

Biology. The aphids live in the leaf sheaths of *Spartina* in coastal saltings and are able to survive submergence of the host. It has also been recorded from *Puccinellia maritima* and *Festuca rubra.* Walker (1848) found sexuales in England in October.

Subgenus *Rungsia* Mimeur, 1933

Rungsia Mimeur, 1933: 104.
Type-species: *Rungsia graminis* Mimeur, 1933
= *Sipha maydis* Passerini, 1860.
Survey: 384.

Dorsal cuticle of apterae not adorned with nodules or denticles between the hairs, but sometimes with very fine spinules (Fig. 307). Distance between eye and antennal base shorter than eye (Fig. 301). First tarsal segm. with 4 or 5 hairs. Cauda broadly rounded (Fig. 312).

Eight species in the world, three species in Scandinavia.

Key to species of *Rungsia*

Apterous and alate viviparous females

1 Dorsum of aptera dark brown or black all over (Fig. 310). Alata with black sclerotic carapace on tergites IV–VII, including siphuncular bases.. 148. *maydis* Passerini
– Dorsum not dark brown or black all over. Alata without black sclerotic carapace on tergites IV–VII. 2
2 (1) Apical segm. of rostrum shorter than 2 × its basal width (Fig. 315). .... 147. *elegans* del Guercio
– Apical segm. of rostrum about twice as long as its basal width (Fig. 314). .... 146. *arenarii* Mordvilko

146. ***Sipha (Rungsia) arenarii*** Mordvilko, 1921
Fig. 314.

Sipha arenarii Mordvilko, 1921: 57. – Survey: 398.

Apterous viviparous female. Greenish yellow to brown with pale longitudinal stripes. Dorsal cuticle without denticles between the long, pointed hairs. Antenna about 0.33 × body; processus terminalis about 2 × Va; segm. III 1.2–1.4 × segm. V. Apical segm. of rostrum about twice as long as its basal width (Fig. 314), about 0.7 × 2sht., about as long as segm. III of rostrum. 2.0–2.4 mm.

Oviparous female. Often rather dark. Abdomen with dark intersegmental stripes. Hind tibiae slightly thickened, with some roundish or irregularly shaped scent plaques on dark basal $^2/_3$.

Apterous male. Slender. Antenna relatively longer than in apterous females; segm. III and IV with numerous small secondary rhinaria all over. Rather small.

Distribution. In Denmark found in NEZ (Humlebæk); in Sweden in Sk., Hall., and Gtl.; in Norway in Bø; in Finland in N, Ka, St, and ObN. – Poland, including the Baltic region, and the USSR, including NW Russia, east to Kazakhstan, W Siberia, and C Asia.

Biology. The species lives on *Elymus arenarius*. Sexuales occur in Denmark in October.

147. ***Sipha (Rungsia) elegans*** del Guercio, 1905
Figs. 301, 315.

Sipha elegans del Guercio, 1905: 137.
Sipha kurdjumovi Mordvilko, 1921: 56.
Sipha agropyrella Hille Ris Lambers, 1939b: 82.
Survey: 398.

Apterous viviparous female. Yellowish brown to brown, with paler longitudinal median stripe. Abdomen with dark intersegmental sclerites. Siphunculi dark. Dorsal cuticle without denticles between the long, pointed hairs. Antenna about 0.4 × body; processus terminalis 1.5–2 × Va; segm. III as long as segm. V or a little shorter. Apical segm. of rostrum shorter than 2 × its width at base (Fig. 315), about 0.6 × 2sht., shorter than segm. III of rostrum. 1.6–2.2 mm.

Alate viviparous female. Abdomen with dark spots and short cross bars not touching the siphuncular bases. Antenna nearly 0.5 × body; segm. III with 4–10 rhinaria.

Oviparous female. Rather dark. Hind tibiae swollen, with many scent plaques.

Apterous male. Black. Slender. Antenna about 0.65 × body; rather thick; secondary rhinaria on segm. III: 40–60, IV: 8–14; antennal hairs rather short. About 1.5 mm.

Distribution. In Denmark found in WJ, NWJ, NEJ, and NEZ, not rare; in Sweden in Sk., Bl., Hall., Sm., Ög., and Upl.; in Norway in Ø; in Finland in N. – Europe and Asia, south to Spain, Yugoslavia, and Turkey, east to W Siberia and C Asia; moderately common in Great Britain; common in N Germany; known from the Baltic region of Poland, but not from NW & W Russia. N America: widespread in the USA and Canada.

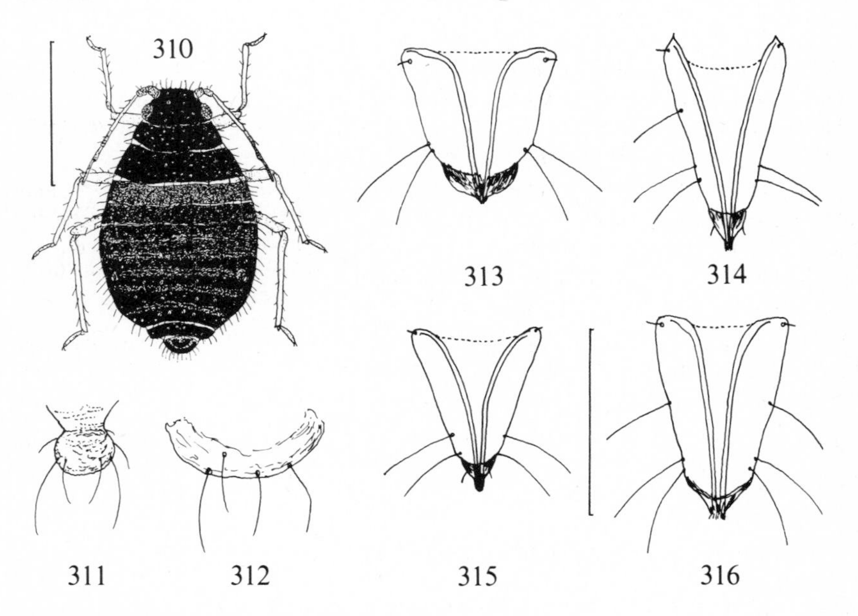

Fig. 310. *Sipha (Rungsia) maydis* Pass. (Scale 1 mm).

Figs. 311, 312. Cauda of *Sipha* s. str. (311) and subgenus *Rungsia* (312).

Figs. 313–316. Apical segm. of rostrum of *Sipha* spp. – 313: *glyceriae* (Kalt.); 314: *arenarii* Mordv.; 315: *elegans* (d. Guerc.); 316: *maydis* Pass. (Scale 0.1 mm).

Biology. The typical host is probably *Agropyrum repens*, but the species can also feed on several other grasses as *Festuca, Hordeum, Arrhenatherum,* and *Triticum*. It is not visited by ants.

148. ***Sipha (Rungsia) maydis*** Passerini, 1860
Plate 4: 11. Figs. 307, 310, 312.

Sipha maydis Passerini, 1860: 38. – Survey: 398.

Apterous viviparous female. Shining dark brown or black. Antennae pale with darker apices. Legs pale. Dorsal cuticle in some places with rows of very fine spinules forming a hardly visible reticulate pattern, most distinctly visible on the posterior part of abdomen. Dorsal hairs long, placed on tubercular, pale bases. Antenna about 0.5 × body; processus terminalis about 1.6 × Va or shorter; segm. III about as long as segm. V. Apical segm. of rostrum about twice as long as its basal width, about 0.75 × 2sht., about as long as segm. III of rostrum. 1.6–2.1 mm.

Alate viviparous female. Abdomen with a solid, black, sclerotic carapace on tergites IV–VII, including the siphuncular bases and the marginal sclerites of tergites VI and VII, as well as transverse dorsal bands on tergites I–III. Ant. segm. III with about 4 rhinaria on the middle part.

Oviparous female. Basal $^4/_5$ of hind tibia dark, thickened, with 80–100 small, apparently double scent plaques. Otherwise much like the apterous viviparous female.

Apterous male. Black. Antennae black except segm. II and the very base of segm. III. Secondary rhinaria on ant. segm. III: 45–50, IV: 8–16, Va: 0.

Distribution. In Denmark found in NEJ; in Sweden known from Sk., Vg., Upl., and Vstm.; recorded from Norway; in Finland known from Sa. – In Europe south to the Mediterranean Sea; rare in Great Britain (Surrey); common in N Germany; known from the Baltic region of Poland; in the USSR north to Minsk, east to W Siberia and C Asia; Turkey, Lebanon, Israel; N Africa, S Africa.

Biology. The species lives at the leaf bases of various grasses, including cereals, e.g. *Zea, Avena,* and *Triticum*. In Sweden rercorded from *Festuca pratensis*, in Britain only from *Arrhenatherum elatius* and *Agropyrum repens*. It is visited by ants.

Note. The sexuales are described on the basis of Tuatay & Remaudière (1964) who found them in Turkey.

Genus *Chaetosiphella* Hille Ris Lambers, 1939

Chaetosiphella Hille Ris Lambers, 1939b: 84.
Type-species: *Sipha berlesei* del Guercio, 1905.
Survey: 134.

Apical segm. of rostrum very long and pointed, longer than 2sht., also longer than ant.

segm. III, stylet-shaped. Ocular tubercles absent. Antenna 5-segmented. Empodial hairs spatulate. Wings narrow, with normal venation. Siphunculus pore-shaped, placed on anterior part of margin of abd. segm. V. Cauda rounded, not constricted.

Three species in the world, one in Scandinavia. One more species is included below, because it has been found in N Germany and possibly will be found also in southern Scandinavia. They live on grasses, with or without attention of ants.

Key to species of *Chaetosiphella*

Apterous viviparous females

1 Dorsal hairs thorn-like. .... 149. *berlesei* (del Guercio)
– Dorsal hairs fan-shaped (Fig. 325). .... *tshernavini* (Mordvilko)

149. *Chaetosiphella berlesei* (del Guercio, 1905)
Plate 4: 14. Figs. 317, 318, 320, 322.

Sipha berlesei del Guercio, 1905: 135. – Survey: 134.

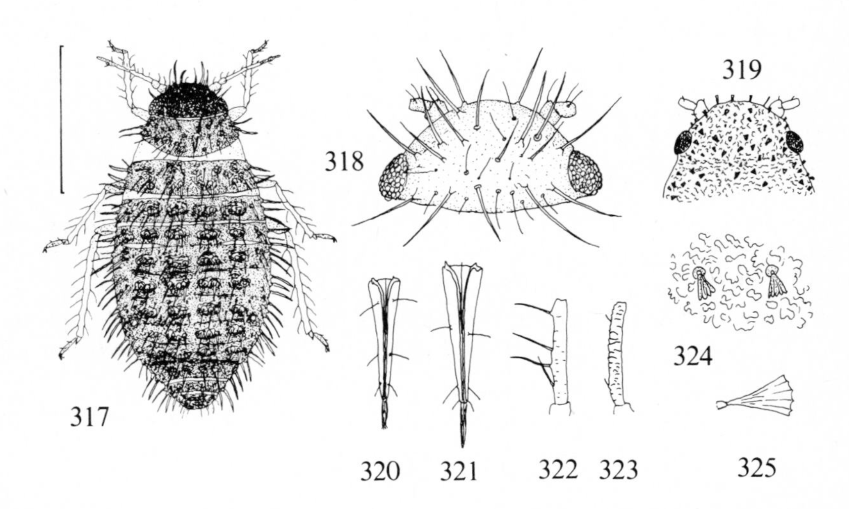

Figs. 317–325. *Chaetosiphella* spp., apt. viv. – 317: habitus of *berlesei* (d. Guerc.); 318: head of *berlesei;* 319: head of *tshernavini* (Mordv.); 320, 321: apical segm. of rostrum of *berlesei* (320) and *tshernavini* (321); 322, 323: ant. segm. III of *berlesei* (322) and *tshernavini* (323); 324: part of dorsal cuticle of *tshernavini;* 325: dorsal body hair of same. (Scale 1 mm for 317). (318 after Richards, 319–324 after Szelegiewicz, 325 after Shaposhnikov, all redrawn; 317 orig.).

Apterous viviparous female. Dull black. Body elongate, slender; abdomen broader than thorax. Dorsum sclerotic, with many thick, pale, thorn-like hairs, some of them long and placed on wart-like bases, but most of them short; the short hairs are absent from nymphs. Antenna about 0.25 × body; processus terminalis shorter than Va. About 2 mm.

Alate viviparous female. Body a little broader than in the apterous female. Abdomen with spinal, pleural, and marginal sclerites, more or less fused into short cross bars on posterior segments. Antenna about 0.4 × body; processus terminalis about as long as Va; segm. III with 4–5 rhinaria.

Oviparous female. Dorsum with cross bars instead of a solid sclerotic shield. Hind tibiae swollen, with numerous 8-shaped scent plaques.

Apterous male. Body slender, almost linear. Dorsum sclerotic, only with long hairs. Antenna longer than 0.5 × body; secondary rhinaria on segm. III: 30–45, IV: 12–22, Va: 0–3.

Distribution. In Denmark found in NWJ and NEJ; in Sweden widespread, from Sk. in the south to T.Lpm. in the north; in Norway known from Bø; in Finland from N. – The Netherlands, Belgium, Germany (not N Germany), Poland (including the Baltic region), Czechoslovakia, Hungary, Italy, W Siberia.

Biology. The aphids live on the uppersides of leaves of *Deschampsia flexuosa* and *Festuca ovina,* mostly in very arid environment, e.g. dune areas. They are able to run very fast, but drop off the plant when disturbed.

Chaetosiphella tshernavini (Mordvilko, 1921)
Figs. 319, 321, 323–325.

Sipha tshernavini Mordvilko, 1921: 57. – Survey: 135.

Apterous viviparous female. Blackish grey. Most characters as in *berlesei,* but dorsal hairs, except those on frons and on margins of abd. tergites VII and VIII, thick, straight, and fan-shaped, with widened apices (Figs. 324, 325). 1.5–1.8 mm.

Distribution. Not yet found in Scandinavia. – Germany (including N Germany, Rostock (F. P. Müller in litt.)), Poland (not the Baltic region), Czechoslovakia, S Russia.

Biology. The species lives on the leaves of *Corynephorus canescens,* and is visited by ants.

				DENMARK												
		N. Germany	G. Britain	SJ	EJ	WJ	NWJ	NEJ	F	LFM	SZ	NWZ	NEZ	B	Sk.	Bl.
Drepanosiphum acerinum (Wlk.)	55		●		●		●	●								
D. aceris Koch	56	●	●	●	●	●	●	●	●	●		●				
D. dixoni H. R. L.			●													
D. oregonensis Gran.		●														
D. platanoidis (Schrk.)	57	●	●	●	●	●	●	●	●	●	●	●	●		●	
Symydobius oblongus (v. Heyd.)	58	●	●	●	●	●	●	●	●		●		●	●	●	●
Clethrobius comes (Wlk.)	59		●		●		●	●	●				●		●	●
Euceraphis betulae (Koch)	60	●	●	●	●	●	●	●	●	●	●	●	●	●	●	
E. punctipennis (Zett.)	61	●	●	●	●	●	●	●	●	●	●	●	●	●		
Phyllaphis fagi (L.).	62	●	●	●	●	●	●	●	●	●	●	●	●	●	●	●
Callipterinella calliptera (Htg.)	63	●	●		●		●	●					●		●	
C. minutissima (Stroyan)	64		●				●								●	
C. tuberculata (v. Heyd.)	65	●	●										●		●	●
Calaphis betulicola (Kalt.)	66	●	●		●			●							●	
C. flava Mord.	67	●	●	●	●	●	●	●	●	●	●		●		●	
Betulaphis brevipilosa Börn.	68	●			●	●	●	●	●	●			●		●	
B. pelei H. R. L.	69															
B. quadrituberculata (Kalt.)	70	●	●	●	●	●	●	●	●	●			●	●	●	●
Monaphis antennata (Kalt.)	71		●												●	
Callaphis juglandis (Goeze)	72	●	●	●	●								●		●	
Chromaphis juglandicola (Kalt.)	73	●	●	●									●		●	
Myzocallis carpini (Koch)	74	●	●	●	●	●	●	●	●	●	●		●	●	●	●
M. castanicola Baker	75	●	●		●		●	●							●	●
M. coryli (Goeze)	76	●	●	●	●	●	●	●	●	●	●		●		●	●
M. myricae (Kalt.)	77	●	●	●	●	●	●	●								●
Tuberculatus (T.) querceus (Kalt.)	78	●	●		●		●						●		●	
T. (Tuberculoides) annulatus (Htg.)	79	●	●	●	●	●	●	●	●	●	●	●	●	●	●	●
T. (T.) borealis (Krzywiec)	80		●										●		●	
T. (T.) neglectus (Krzywiec)	81		●		●											
Pterocallis albidus Börn.	82					●	●								●	
P. alni (DeGeer)	83	●	●	●	●	●	●	●		●	●	●	●	●	●	●
P. maculatus (v. Heyd.)	84		●	●			●									
Ctenocallis setosus (Kalt.)	85	●	●		●											
Tinocallis platani (Kalt.)	86	●	●				●						●		●	
T. saltans (Nevsky)	87														●	
Eucallipterus tiliae (L.)	88	●	●	●	●	●	●	●	●	●	●		●		●	
Therioaphis (T.) luteola (Börn.)	89	●	●						●		●		●		●	
T. (T.) ononidis (Kalt.)	90	●	●		●		●	●	●						●	

SWEDEN

	Hall.	Sm.	Öl.	Gtl.	G. Sand.	Ög.	Vg.	Boh.	Dlsl.	Nrk.	Sdm.	Upl.	Vstm.	Vrm.	Dlr.	Gstr.	Hls.	Med.	Hrj.	Jmt.	Ång.	Vb.	Nb.	Ås. Lpm.	Ly. Lpm.	P. Lpm.	Lu. Lpm.	T. Lpm.
55																												
56						●						●																
57	●	●	●	●			●			●	●	●	●															
58	●	●	●	●		●	●	●	●			●	●	●	●		●		●	●	●			●		●	●	●
59	●						●	●				●			●		●		●		●		●				●	●
60		●	●	●		●				●			●	●				●										
61	●	●				●											●					●	●		●		●	●
62	●	●	●	●		●	●				●	●	●															
63	●		●									●						●			●	●	●					
64																												
65	●	●	●					●			●	●																
66		●	●	●		●	●		●			●	●		●			●				●						
67	●	●		●		●	●		●		●	●	●	●	●		●	●		●		●					●	
68		●				●	●					●		●					●	●								
69																												●
70											●	●														●	●	
71		●				●	●					●			●													
72																												
73			●	●																								
74			●	●						●		●																
75												●																
76	●		●			●	●	●	●	●	●	●			●													
77	●	●		●					●		●	●	●	●	●		●					●						
78	●		●																									
79	●	●	●	●		●	●	●	●	●	●	●	●	●	●							●						
80																												
81																												
82																						●						
83	●	●	●	●						●	●	●			●	●	●		●			●	●					
84			●									●																
85																												
86				●																								
87																												
88		●	●	●			●			●	●	●	●	●	●		●	●				●						
89			●			●	●																					
90			●																									

		NORWAY														
		Ø + AK	HE (s + n)	O (s + n)	B (ø + v)	VE	TE (y + i)	AA (y + i)	VA (y + i)	R (y + i)	HO (y + i)	SF (y + i)	MR (y + i)	ST (y + i)	NT (y + i)	Ns (y + i)
Drepanosiphum acerinum (Wlk.)	55															
D. aceris Koch	56															
D. dixoni H. R. L.																
D. oregonensis Gran.																
D. platanoidis (Schrk.)	57	●		●	●						●					
Symydobius oblongus (v. Heyd.)	58	●		●	●			●	●	●	●					
Clethrobius comes (Wlk.)	59	●	●		●						●		●			
Euceraphis betulae (Koch)	60											●				
E. punctipennis (Zett.)	61			●												
Phyllaphis fagi (L.).	62	●				●			●		●	●				
Callipterinella calliptera (Htg.)	63											●				
C. minutissima (Stroyan)	64															
C. tuberculata (v. Heyd.)	65	●	●	●	●			●				●				
Calaphis betulicola (Kalt.)	66	●														
C. flava Mord.	67	●		●						●	●					
Betulaphis brevipilosa Börn.	68				●							●				
B. pelei H. R. L.	69			●	●						●		●			
B. quadrituberculata (Kalt.)	70			●	●			●			●	●				
Monaphis antennata (Kalt.)	71															
Callaphis juglandis (Goeze)	72															
Chromaphis juglandicola (Kalt.)	73															
Myzocallis carpini (Koch)	74	●														
M. castanicola Baker	75															
M. coryli (Goeze)	76	●		●	●					●	●	●				
M. myricae (Kalt.)	77	●					●	●			●		●			
Tuberculatus (T.) querceus (Kalt.)	78	●									●					
T. (Tuberculoides) annulatus (Htg.)	79	●			●	●	●	●	●	●	●					
T. (T.) borealis (Krzywiec)	80															
T. (T.) neglectus (Krzywiec)	81					**N**	**O**	**R**	**W**	**A**	**Y**					
Pterocallis albidus Börn.	82															
P. alni (DeGeer)	83	●	●	●	●						●	●			●	
P. maculatus (v. Heyd.)	84															
Ctenocallis setosus (Kalt.)	85															
Tinocallis platani (Kalt.)	86	●		●												
T. saltans (Nevsky)	87															
Eucallipterus tiliae (L.)	88	●			●					●		●			●	
Therioaphis (T.) luteola (Börn.)	89															
T. (T.) ononidis (Kalt.)	90															

	FINLAND																								USSR		
	Nn (ø + v)	TR (y + i)	F (v + i)	F (n + ø)	Al	Ab	N	Ka	St	Ta	Sa	Oa	Tb	Sb	Kb	Om	Ok	Ob S	Ob N	Ks	LkW	LkE	Le	Li	Vib	Kr	Lr
55																											
56																											
57							●																				
58					●	●	●			●	●	●		●	●		●		●					●		●	
59						●	●			●	●																
60						●	●			●							●							●			
61												●							●			●		●			
62							●																				
63											●																
64																											
65		●			●		●			●	●		●													●	
66							●										●		●								
67							●			●	●	●					●										
68							●			●	●	●								●		●		●			
69																	●						●				
70							●			●							●							●			
71																											
72																											
73																											
74																											
75																											
76						●	●			●	●																
77										●	●																
78																											
79					●	●	●			●	●							●									
80																											
81																											
82										●							●										
83		●				●	●		●									●									
84										●	●																
85																											
86							●			●																	
87																											
88					●	●	●			●	●	●	●	●			●	●									
89							●																				
90																											

				DENMARK												
		N. Germany	G. Britain	SJ	EJ	WJ	NWJ	NEJ	F	LFM	SZ	NWZ	NEZ	B	Sk.	Bl.
T. (T.) riehmi (Börn.)	91	●	●		●		●						●		●	
T. (T.) subalba (Börn.)	92															
T. (T.) trifolii (Monell)	93	●	●	●	●	●	●	●			●		●		●	
T. (Rhizoberlesia) brachytricha	94															
Thripsaphis (T.) ballii caespitosa	95															
T. (T.) caricicola (Mord.)	96														●	
T. (Trichocallis) caricis (Mord.)	97	●	●												●	
T. (T.) cyperi (Wlk.)	98		●		●	●									●	●
T. (T.) ossiannilssoni H. R. L.	99	●														
T. (T.) verrucosa Gill.	100		●					●							●	
T. (T.) vibei arctica H. R. L.	101															
T. (Larvaphis) brevicornis Oss.	102															
Subsaltusaphis aquatilis (Oss.)	103															
S. flava (H. R. L.)	104	●	●			●	●	●								●
S. intermedia (H. R. L.)																
S. lambersi (Quedn.)	105														●	
S. maritima (H. R. L.)																
S. ornata (Theob.)	106	●	●												●	
S. pallida (H. R. L.	107						●									
S. paniceae (Quedn.)	108		●												●	
S. picta (H. R. L.)	109	●	●												●	
S. rossneri (Börn.)	110		●			●									●	
Saltusaphis lasiocarpae (Oss.)	111															
S. scirpus Theobald																
Nevskyella fungifera (Oss.)	112															
Iziphya austriaca Börn.	113														●	
I. bufo (Wlk.)	114	●	●			●	●	●					●		●	●
I. ingegardae H. R. L.	115															
I. leegei Börn.	116		●				●						●		●	
I. memorialis Börn.	117															
Periphyllus acericola (Wlk.)	118	●	●	●	●	●	●	●	●	●		●	●		●	
P. aceris (L.)	119		●				●								●	●
P. californiensis (Shinji)			●													
P. coracinus (Koch)	120															
P. hirticornis (Wlk.)	121		●							●			●			
P. lyropictus (Kessler)	122	●	●	●	●		●								●	
P. obscurus Mamontova			●													
P. singeri (Börn.)																

SWEDEN

	Hall.	Sm.	Öl.	Gtl.	G. Sand.	Ög.	Vg.	Boh.	Dlsl.	Nrk.	Sdm.	Upl.	Vstm.	Vrm.	Dlr.	Gstr.	Hls.	Med.	Hrj.	Jmt.	Äng.	Vb.	Nb.	Ås. Lpm.	Ly. Lpm.	P. Lpm.	Lu. Lpm.	T. Lpm.
91												●	●															
92						●	●			●		●	●	●	●					●								
93			●	●		●		●		●		●	●		●			●		●		●						
94		●	●			●						●	●	●						●								
95			●									●																●
96			●									●	●		●													
97	●						●	●		●	●	●	●	●	●		●	●				●						
98		●				●			●	●		●	●	●	●		●	●	●	●	●	●	●				●	●
99			●								●	●																
100	●		●			●	●	●	●	●	●	●	●	●	●		●	●	●	●								●
101																											●	
102												●			●													
103															●			●									●	
104	●	●	●			●	●				●	●	●	●			●	●		●			●					
105										●		●																
106																												
107			●	●								●																
108			●	●								●	●							●								
109			●			●			●			●							●									
110		●				●	●			●		●	●							●						●		
111												●									●	●	●					
112												●																
113											●	●	●	●								●						
114	●		●		●								●	●				●										
115											●	●	●	●														
116			●	●				●				●	●				●											
117												●	●															
118																												
119	●	●	●								●	●			●		●											
120		●										●																
121																												
122		●	●						●	●		●			●		●	●			●	●	●					

		NORWAY														
		Ø + AK	HE (s + n)	O (s + n)	B (ø + v)	VE	TE (y + i)	AA (y + i)	VA (y + i)	R (y + i)	HO (y + i)	SF (y + i)	MR (y + i)	ST (y + i)	NT (y + i)	Ns (y + i)
T. (T.) riehmi (Börn.)	91															
T. (T.) subalba (Börn.)	92															
T. (T.) trifolii (Monell)	93	●														
T. (Rhizoberlesia) brachytricha	94	●		●							●					
Thripsaphis (T.) ballii caespitosa	95															
T. (T.) caricicola (Mord.)	96															
T. (Trichocallis) caricis (Mord.)	97			●	●						●	●				
T. (T.) cyperi (Wlk.)	98			●							●		●			
T. (T.) ossiannilssoni H. R. L.	99															
T. (T.) verrucosa Gill.	100										●					
T. (T.) vibei arctica H. R. L.	101															
T. (Larvaphis) brevicornis Oss.	102															
Subsaltusaphis aquatilis (Oss.)	103															
S. flava (H. R. L.)	104			●							●					
S. intermedia (H. R. L.)																
S. lambersi (Quedn.)	105															
S. maritima (H. R. L.)																
S. ornata (Theob.)	106															
S. pallida (H. R. L.	107															
S. paniceae (Quedn.)	108															
S. picta (H. R. L.)	109															
S. rossneri (Börn.)	110		●													
Saltusaphis lasiocarpae (Oss.)	111															
S. scirpus Theobald																
Nevskyella fungifera (Oss.)	112															
Iziphya austriaca Börn.	113															
I. bufo (Wlk.)	114															
I. ingegardae H. R. L.	115															
I. leegei Börn.	116										●					
I. memorialis Börn.	117															
Periphyllus acericola (Wlk.)	118					**N**	**O**	**R**	**W**	**A**	**Y**					
P. aceris (L.)	119															
P. californiensis (Shinji)																
P. coracinus (Koch)	120															
P. hirticornis (Wlk.)	121															
P. lyropictus (Kessler)	122															
P. obscurus Mamontova																
P. singeri (Börn.)																

					FINLAND																			USSR		
	Nn (ø + v)	TR (y + i)	F (v + i)	F (n + ø)	Al	Ab	N	Ka	St	Ta	Sa	Oa	Tb	Sb	Kb	Om	Ok	Ob S	Ob N	Ks	LkW	LkE	Le	Li	Vib	Kr	Lr
91																											
92							●																				
93							●																				
94																											
95																											
96																											
97							●				●						●		●								
98																	●		●								
99																											
100																	●										
101																											
102																											
103																											
104																			●								
105																											
106																											
107																											
108																											
109											●																
110																	●										
111																											
112																											
113							●										●										
114																											
115																											
116																											
117																											
118																											
119					●	●	●		●	●	●	●				●		●	●						●		
120																											
121																											
122											●																

				DENMARK												
		N. Germany	G. Britain	SJ	EJ	WJ	NWJ	NEJ	F	LFM	SZ	NWZ	NEZ	B	Sk.	Bl.
P. testudinaceus (Fern.)	123	●	●	●	●	●	●	●	●	●	●	●	●	●	●	●
Chaitophorus capreae (Mosley)	124	●	●	●	●	●	●	●	●				●		●	
C. horii beuthani (Börn.)	125	●	●	●	●		●	●	●		●		●		●	
C. lapponum Oss.	126															
C. leucomelas Koch	127	●	●		●		●	●		●		●	●		●	
C. parvus H. R. L.	128							●							●	
C. pentandrinus Oss.	129															
C. populeti (Panz.)	130	●	●		●	●	●	●	●	●			●		●	●
C. populialbae B. d. F.	131	●	●				●						●		●	
C. ramicola (Börn.)	132															
C. salicti (Schrk.)	133	●	●				●	●					●		●	●
C. salijaponicus niger Mord.	134	●	●												●	
C. tremulae Koch	135	●	●		●	●	●	●							●	●
C. truncatus (Hausm.)	136	●	●										●			
C. vitellinae (Schrk.)	137	●	●										●		●	
Carisosipha paniculatae Börn.	138	●	●		●	●	●						●		●	
Laingia psammae Theobald	139	●	●			●	●	●							●	
Atheroides brevicornis Laing	140		●													
A. doncasteri Oss.	141															
A. hirtellus Hal.	142		●													
A. serrulatus Hal.	143	●	●		●	●	●	●					●		●	●
Sipha (S.) glyceriae (Kalt.)	144	●	●		●	●	●	●			●		●		●	●
S. (S.) littoralis (Wlk.)	145	●	●	●												
S. (Rungsia) arenarii Mord.	146												●		●	
S. (R.) elegans d. Guerc.	147	●	●			●	●	●					●		●	●
S. (R.) maydis Pass.	148	●	●					●							●	
Chaetosiphella berlesei (d. Guerc.)	149						●	●							●	●
C. tshernavini (Mord.)		●														

SWEDEN

	Hall.	Sm.	Öl.	Gtl.	G. Sand.	Ög.	Vg.	Boh.	Dlsl.	Nrk.	Sdm.	Upl.	Vstm.	Vrm.	Dlr.	Gstr.	Hls.	Med.	Hrj.	Jmt.	Ång.	Vb.	Nb.	Ås. Lpm.	Ly. Lpm.	P. Lpm.	Lu. Lpm.	T. Lpm.
123	●	●	●	●			●	●		●	●	●	●	●	●						●							
124		●	●			●						●					●	●										
125												●																
126															●				●			●	●			●	●	
127		●	●	●		●		●				●																
128								●																				
129						●						●																
130	●	●	●			●	●	●		●	●	●	●	●	●			●				●	●					●
131												●															●	●
132												●																
133		●	●				●					●			●			●										
134												●																
135		●	●	●		●	●		●	●	●	●	●	●	●		●	●		●		●	●			●		
136																												
137																												
138			●	●		●				●	●	●															●	
139	●			●					●	●		●	●	●														
140								●																				
141										●	●	●																
142		●								●	●																	
143	●	●	●	●		●	●	●	●	●		●	●	●	●		●	●	●	●	●	●	●		●		●	
144			●	●		●	●	●	●			●	●	●			●											
145																							●					
146	●			●																								
147	●	●				●						●																
148		●					●					●	●															
149								●		●		●	●		●			●		●	●	●				●		●

		NORWAY														
		Ø + AK	HE (s + n)	O (s + n)	B (ø + v)	VE	TE (y + i)	AA (y + i)	VA (y + i)	R (y + i)	HO (y + i)	SF (y + i)	MR (y + i)	ST (y + i)	NT (y + i)	Ns (y + i)
P. testudinaceus (Fern.)	123	●								●	●					
Chaitophorus capreae (Mosley)	124			●												●
C. horii beuthani (Börn.)	125															
C. lapponum Oss.	126															
C. leucomelas Koch	127				●											
C. parvus H. R. L.	128															
C. pentandrinus Oss.	129															
C. populeti (Panz.)	130	●			●		●				●		●			
C. populialbae B. d. F.	131	●														
C. ramicola (Börn.)	132															
C. salicti (Schrk.)	133															
C. salijaponicus niger Mord.	134				●						●					
C. tremulae Koch	135	●			●					●	●				●	●
C. truncatus (Hausm.)	136			●												
C. vitellinae (Schrk.)	137															
Carisosipha paniculatae Börn.	138	●														
Laingia psammae Theobald	139															
Atheroides brevicornis Laing	140										●					
A. doncasteri Oss.	141															
A. hirtellus Hal.	142															
A. serrulatus Hal.	143	●								●	●					
Sipha (S.). glyceriae (Kalt.)	144		●							●	●	●				
S. (S.) littoralis (Wlk.)	145															
S. (Rungsia) arenarii Mord.	146				●											
S. (R.) elegans d. Guerc.	147	●														
S. (R.) maydis Pass.	148					**N**	**O**	**R**	**W**	**A**	**Y**					
Chaetosiphella berlesei (d. Guerc.)	149				●											
C. tshernavini (Mord.)																

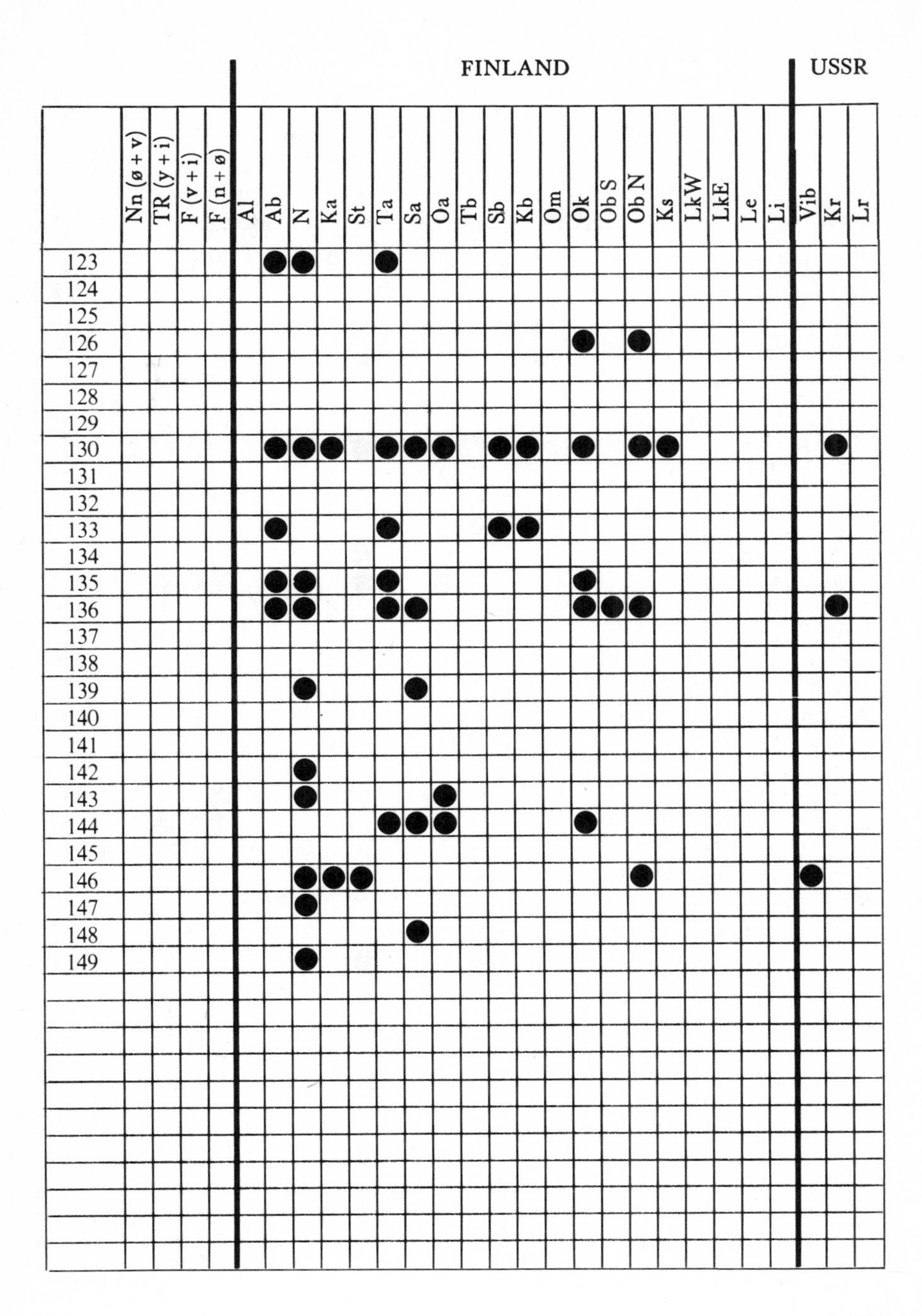

					FINLAND																				USSR		
	Nn (ø + v)	TR (y + i)	F (v + i)	F (n + ø)	Al	Ab	N	Ka	St	Ta	Sa	Oa	Tb	Sb	Kb	Om	Ok	Ob S	Ob N	Ks	LkW	LkE	Le	Li	Vib	Kr	Lr
123						●	●			●																	
124																											
125																											
126																	●		●								
127																											
128																											
129																											
130						●	●	●		●	●	●		●	●		●		●	●						●	
131																											
132																											
133						●				●				●	●												
134																											
135						●	●			●							●										
136						●	●			●	●						●	●	●							●	
137																											
138																											
139							●				●																
140																											
141																											
142							●																				
143							●					●															
144										●	●	●					●										
145																											
146							●	●	●										●						●		
147							●																				
148											●																
149							●																				

Literature

(continued from vol. 9)

Baker, A. C., 1917: Eastern aphids, new or little known. II. – J. econ. Ent., 10: 420–433.

Bissell, T. L., 1978: Aphids on Juglandaceae in North America. – Misc. Publ. Maryland Agric. Exp. Sta., No. 911, 78 pp.

Blackmann, R. L., 1976: Cytogenetics of two species of *Euceraphis* (Homoptera, Aphididae). – Cromosoma, 56: 393–408.

– , 1977: The existence of two species of *Euceraphis* (Homoptera: Aphididae) on birch in western Europe, and a key to European and North American species of the genus. – Syst. Ent., 2: 1–8.

Börner, C., 1939: Neue Gattungen und Arten der mitteleuropäischen Aphidenfauna. – Arb. phys. u. angew. Ent., 6: 75–83.

– , 1940: Neue Blattläuse aus Mitteleuropa. – Selbstverlag, Naumburg, pp. 1–4.

– , 1949: Kleine Beiträge zur Monographie der europäischen Blattläuse. – Beitr. taxon. Zool., 1: 44–62.

Buckton, G. B., 1900: Notes on two new species of aphids. – Indian Mus. Notes, 4: 277–278.

Carver, M., 1978: The scientific nomenclature of the spotted alfalfa aphid (Homoptera: Aphididae). – J. Aust. ent. Soc., 17: 287–288.

Cholodkovsky, N., 1896–1911: Aphidologische Mitteilungen. – Zool. Anz., 19 (1896): 508–513, 20 (1897): 145–147, 22 (1899): 468–477, 24 (1901): 292–296, 26 (1903): 258–263, 27 (1903): 118–119, 27 (1904): 476–479, 32 (1908): 687–693), 35 (1910): 279–284, 37 (1911): 172–178.

Dickson, R. C., 1959: On the identity of the spotted alfalfa aphid in North America. – Ann. ent. Soc. Am., 52: 63–68.

Dixon, A. F. G., 1976: Factors determining the distribution of sycamore aphids on sycamore leaves during summer. – Ecol. Ent., 1: 275–278.

– & Logan, M., 1973: Leaf size and availability of space to the sycamore aphid *Drepanosiphum platanoides*. – Oikos, 24: 58–63.

– & McKay, S., 1970: Aggregation in the sycamore aphid *Drepanosiphum platanoides* (Schrk.) (Hemiptera: Aphididae) and its relevance to the regulation of population growth. – J. Anim. Ecol., 39: 439–454.

Essig, E. O. & Abernathy, F., 1952: The aphid genus *Periphyllus*. – Univ. of California Press, 166 pp.

– & Kuwana, S. S., 1918: Some Japanese Aphididae. – Proc. Calif. Acad. Sci. (Ser. 4), 8: 35–112.

Fernie, T. P., 1852: On a new insect, a parasite of sycamore. – Naturalist (Morris), 3: 265.

Forbes, A. R., Frazer, B. D. & Chan, C.–K., 1974: The aphids (Homoptera: Aphididae) of British Colombia. 3. Additions and corrections. – J. ent. Soc. Brit. Columbia, 71: 43–49.

Frisch, J. L., 1734: Beschreibung von allerley Insecten in Deutschland, nebst nützlichen Anmerkungen und nöthigen Abbildungen. 1. Aufl. Berlin. Tom. XI.
Gillette, C. P., 1907–08: New species of Colorado Aphididae, with notes upon their life-habits. – Can. Ent., 39 (1907): 389–396, 40 (1908): 17–20, 61–68.
– , 1917: Two new aphid genera and some new species. – Ibid. 49: 193–199.
Glendenning, R., 1926: Some new aphids from British Columbia. – Ibid. 58: 95–98.
Goeze, J. A., 1778: Entomologische Beiträge zu des Ritters Linné zwölfter Ausgabe des Natursystems, 2: 286–318.
Granovsky, A. A., 1939: Three new species of Aphiidae (Homoptera). – Proc. ent. Soc. Wash., 41: 143–154.
Guercio, G. del, 1905: Contribuzione alle conoscenza delle *Sipha* Pass. ed alla loro posizione nella Famiglia degli Afidi. – Redia, 2: 127–153.
– , 1915: Ulteriori richerche sullo stremenzimento o incappuciamento del Trifoglio. – Ibid. 10: 235–301.
Haliday, A. H., 1839: New British insects indicated in Mr. Curtis's Guide. – Ann. nat. Hist., 2: 183–190.
Heie, O. E., 1972c: Bladlus på birk i Danmark (Homoptera, Aphidoidea) (Aphids on birch in Denmark). – Ent. Meddr, 40: 81–105.
– , 1980: The Aphidoidea (Hemiptera) of Fennoscandia and Denmark. I. General part. The families Mindaridae, Hormaphididae, Thelaxidae, Anoeciidae, and Pemphigidae. – Fauna ent. scand., 9: 1–236.
Hille Ris Lambers, D., 1935: New Central European Aphididae. – Arb. Morphol. Taxonom. Entomol. Berlin-Dahlem, 2: 52–55.
– , 1952b: New aphids from Sweden. – Opusc. ent. 17: 51–58.
– , 1956: On aphids from the Netherlands with descriptions of new species (Aphididae, Homoptera). – Tijdschr. Ent., 98: 229–249.
– , 1963: Note on *Drepanosiphum acerinum* (Walker, 1848). – Entomologist, 96: 248.
– , 1966: New and little known members of the aphid fauna of Italy (Hymoptera, Aphididae). – Boll. Zool. agr. Bachic. s. II, 8: 1–32.
– , 1971: Two new taxa of *Drepanosiphum* Koch, 1855 (Homoptera, Aphididae) with a key to species. – Ent. Ber., 31: 72–79.
– , 1974a: New species of *Tuberculatus* Mordvilko, 1894 (Homoptera, Aphididae), with a key to species and some critical notes. – Boll. Zool. agr. Bachic., s. II, 11: 21–82.
– , 1974b: On American aphids with descriptions of a new genus and some new species (Homoptera, Aphididae). – Tijdschr. Ent., 117: 103–155.
– , & Bosch, R. v. d., 1964: On the genus *Therioaphis* Walker, 1870, with descriptions of new species (Homoptera, Aphididae). – Zool. Verh., Leiden, 68: 1–47.
– & Stroyan, H. L. G., 1975: Proposed use of the plenary powers to designate a type-species for the nominal genus *Chaitophorus* C. L. Koch, 1854, a genus based upon a misidentified type-species (class Insecta, order Hemiptera). Z. N. (S.) 1003. – Bull. zool. Nomencl., 32: 141–142.
Hoeven, J. v. d., 1863: Over een klein Hemipterum, dat op de bladen van verschilende soorten van *Acer* gevonden wordt. – Tijdschr. Ent., 6: 1–7.

Holman, J. & Pintera, A., 1977: Aphidodea. – Enumeratio insectorum Bohemoslovakiae. Acta faun. ent. Mus. Nat. Pragae, 15, Sppl. 4: 101–116.
Kessler, H. F., 1886: Die Entwickelungs- und Lebensgeschichte von *Chaitophorus aceris* Koch, *Chaitophorus testudinatus* Thornton und *Chaitophorus lyropictus* Kessler. – N. Acta Acad. Leop. (Halle), 51: 151–179.
Kirkaldy, G. W., 1905: Catalogue of the genera of the hemipterous family Aphidae, with their typical species, together with a list of the species described as new from 1885 to 1905. – Can. Ent., 37: 414–420.
Klodnitzki, J. I., 1924: Novyj vid i rod tli iz okrestnostej Kieva. – Trudy IV vseros. entom.-fitop. sezda v. Moskve v 1922 g. Leningrad. 4: 61–63.
Krzywiec, D., 1966: A new species of *Tuberculoides* v. d. Goot from Poland (Homoptera, Aphidina). – Bull. de l'Acad. Polonaise des Sciences, Cl. II, 13 (1965): 595–600.
– , 1971: *Tuberculoides borealis* sp. n., a new species of aphid from Poland (Homoptera, Aphidoidea). – Ibid. 19: 327–333.
Laing, F., 1920: On the genus *Atheroides* Haliday. – Entomologist's mon. Mag., 6: 38–45.
Mamontova, V. A., 1955: Dendrophilous aphids of the Ukraine. – Acad. Sciences, Ukrainian SSR, Kiev, 1955: 1–91 (in Russian).
– , 1963: New data on the aphid fauna (Homoptera, Aphidoidea) of Ukraine. – Acad. Sci. Ukrainian SSR, Pratsi Inst. Zool., 19: 11–40 (in Russian).
Matsumura, S., 1919: New species and genera of Callipterinae (Aphididae) of Japan. – Trans. Sapporo Nat. Hist. Soc., 7: 99–115.
Mimeur, J.-M., 1933: Aphididae du Maroc. (Deuxième note). – Bull. Soc. Sci. Nat. Maroc, 13: 104–108.
Mordvilko, A. K., 1894: K faune i anatomii sem. Aphididae Privislanskogo Kraja. – Vars. Univ. Izv., 1894 (6–9): 1–112.
– , 1914: Faune de la Russie et des pays limitrophes. Insectes Hémiptères 1. Aphidoidea. – Livr. 1, 1914, CLXIV, pp. 1–236 (in Russian).
– , 1921: Zlakovye tli (Aphidodea). – Izv. Petrogr. obl. St. Zašč. Rast. 3 (3): 1–72.
– , 1929: Previous works on URSS aphids and of limitograph area. – Tr. Prikl. Ent. and Inst. Opiti Agron., 14: 1–100.
Mosley, O., 1841: Aphides. – Gard. Chron., 1: 628, 747–748, 827–828.
Nevsky, V. P., 1929a: Aphids of Central Asia. – Uzbekistan Plant Protect. Exp. Sta., 16: 1–425.
– , 1929b: The plant lice of Middle-Asia. III. – Zool. Anz., 82: 197–228.
Ossiannilsson, F., 1953: Three new Swedish aphids (Hem., Hom.). With description of a new genus and a new subgenus. – Opusc. ent., 18: 233–238.
– , 1954a: *Nevskyella* n. n. for *Nevskya* Ossiannilsson, 1953 (Hem., Aphididae). – Ibid. 19: 1.
– , 1954b: Four new Swedish aphids (Hemiptera, Homoptera). With description of a new genus. – Ent. Tidskr., 75: 117–127.
– , 1955: A new European *Atheroides* (Hem., Hom., Aphid.). With synonymic notes on

Atheroides hirtellus Hal. - Ibid. 76: 127–130.
Panzer, G. W. F., 1793–1823: Faunae Insectorum Germaniae Initia. Deutschlands Insekten.
Quednau, F. W., 1953: Kleine Beiträge zur Kenntnis europäischer Blattläuse. - Zool. Anz., 150: 223–229.
- , 1954: Monographie der mitteleuropäischen Callaphididae (Zierläuse (Homoptera, Aphidina)) unter besonderer Berücksichtigung des ersten Jugendstadiums. I. - Mitt. Biol. Zentralanst. f. Land- u. Forstwirtsch. Berlin-Dahlem, 78, 52 pp.
- , 1954b: Kleine Mitteilungen über Chaitophoriden (Homoptera, Aphidina). - Zool. Anz. 152: 308–312.
- , 1966: A list of aphids from Quebec with descriptions of two new species (Homoptera: Aphidoidea). - Can. Ent. 98: 415–430.
- , 1979: A list of Drepanosiphine aphids from the Democratic People's Republic of Korea with taxonomic notes and descriptions of new species (Homoptera). - Annls zool., 34: 501–525.
Richards, W. R., 1965: The Callaphidini of Canada (Homoptera: Aphididae). - Mem. Ent. Soc. Canada, 44: 1–149.
- , 1967: A review of the *Tinocallis* of the world (Homoptera: Aphididae). - Can. Ent. 99: 536–553.
- , 1968a: A revision of the world fauna of *Tuberculatus*, with descriptions of two new species from China (Homoptera: Aphidae). - Can. Ent., 100: 561–596.
- , 1968b: A synopsis of the world fauna of *Myzocallis* (Homoptera: Aphididae). - Mem. Ent. Soc. Canada, 57: 1–76.
- , 1971: A synopsis of the world fauna of Saltusaphidinae or sedge aphids (Homoptera: Aphididae). - Ibid. 80: 1–97.
- , 1972: The Chaitophorinae of Canada (Homoptera: Aphididae). - Ibid. 87: 1–109.
Rupais, A. A., 1969: Atlas of the Baltic dendrofilous plantlice. - Publishing House Zinatne, Riga, 361 pp.
Schmutterer, H., 1952: Zur Kenntnis der Buchenblattlaus, *Phyllaphis fagi* (L.) (Homoptera, Aphidoidea), einer wichtigen Honigtauerzeugerin auf Buche. - Anz. Schädlingskunde, 25: 1–5.
Schrank, F. v. P., 1901: Fauna Boica. Durchgedachte Geschichte der in Baiern einheimischen und zahmen Thiere. Ingolstadt. Aphiden: 2: 102–140.
Shinji, G. O., 1917: New aphids from California (Hem., Hom.). - Ent. News, 28: 61–64.
Stroyan, H. L. G., 1953: A new British species of *Calaphis* Walsh (Hem., Aphididae). - Entomologist's mon. Mag., 89: 13–16.
Szelegiewicz, H., 1961: Die polnischen Arten der Gattung *Chaitophorus* Koch s. lat. (Homoptera, Aphididae). - Annls zool., 19: 229–352.
- , 1974: Materially do poznania mszyc (Homoptera, Aphididea) Polski. II. Rodzina Chaitophoridae. - Fragm. faun., 19: 285–317.
- , 1976: Aphid species (Homoptera, Aphidoidea) new to the Polish fauna. - Annls zool., 33: 217–225.
Takahashi, R., 1939: Some Aphididae from Hokkaido (Hemiptera). - Insecta Matsumurana, 13: 114–128.

Theobald, F. V., 1915: African Aphididae. Part II. – Bull. ent. Res., 6: 103–153.
– , 1922: A new aphid genus and species found in England. – Ibid. 12: 429–430.
Thornton, J., 1852: *Phillophorus testudinatus*. – Proc. ent. Soc. Lond. (N. 5), 2: 78.
Walker, F., 1848b: Descriptions of Aphides. – Ann. Mag. nat. Hist. (2), 1: 249–260, 328–345, 443–454, 2: 43–48, 95–109, 190–203, 421–431.
– , 1870: Notes on Aphides. – Zoologist, 5: 1996–2001.
Walsh, B. D., 1863: On the genera of Aphidae found in the United States. – Proc. ent. Soc. Philad., 1: 294–311.

Index

Synonyms are given in italics. The number in bold refers to the main treatment of the taxon.

Author's address:
Ole E. Heie
Institute of Biology
The Royal Danish School of Educational Studies
Emdrupvej 101, DK-2400 Copenhagen NV, Denmark